KEYS TO THE FAMILIES OF BRIT...

By LAWRENCE M. JONES-WALTERS
Nature Conservancy Council, Northminster House, Peterborough PE1 1UA

ABSTRACT

Two identification keys to families of spiders occurring in Britain are provided. The first is presented in a conventional dichotomous format; the second as a tabular guide which incorporates behavioural and ecological characters as well as morphological features. Sections on spider morphology, biology, ecology and a glossary are included.

CONTENTS

INTRODUCTION 365
    How to recognise spiders 365
    Biology and Ecology. 367
    Finding and collecting 370
    Examining 376
    Keeping and preserving 377
    Spider conservation 377
    Collecting Code 378
GLOSSARY 380
SYSTEMATIC LIST 386
HOW TO USE THE KEYS 388
REFERENCES 389
DICHOTOMOUS KEY 390
LATERAL KEY 420

INTRODUCTION

How to recognise spiders

THE SPIDERS CAN be distinguished from all other invertebrates by the following four characters:
1. Spiders have eight walking legs (a feature which they share with other members of the Class Arachnida—see Fig. 1).
2. They have no external segmentation on the abdomen.
3. Their fused head and thorax (the cephalothorax) is separated from the abdomen by a very narrow "waist", the pedicel.
4. They have small appendages, the spinnerets, at the hind end of their abdomen.

Their eight walking legs distinguish spiders from insects (which only have six) but it is possible that they may be confused with their close relatives the harvestmen. Harvestmen have no constriction between their cephalothorax and abdomen, and closer examination reveals their abdomen to be clearly segmented. Figure 1 shows a spider and its other British arachnid relatives; (the scorpion illustrated is *Euscorpius flavicaudis* which is common in southern Europe and has established thriving colonies at a number of British localities, mainly sea ports. It is, however, very unlikely to be found).

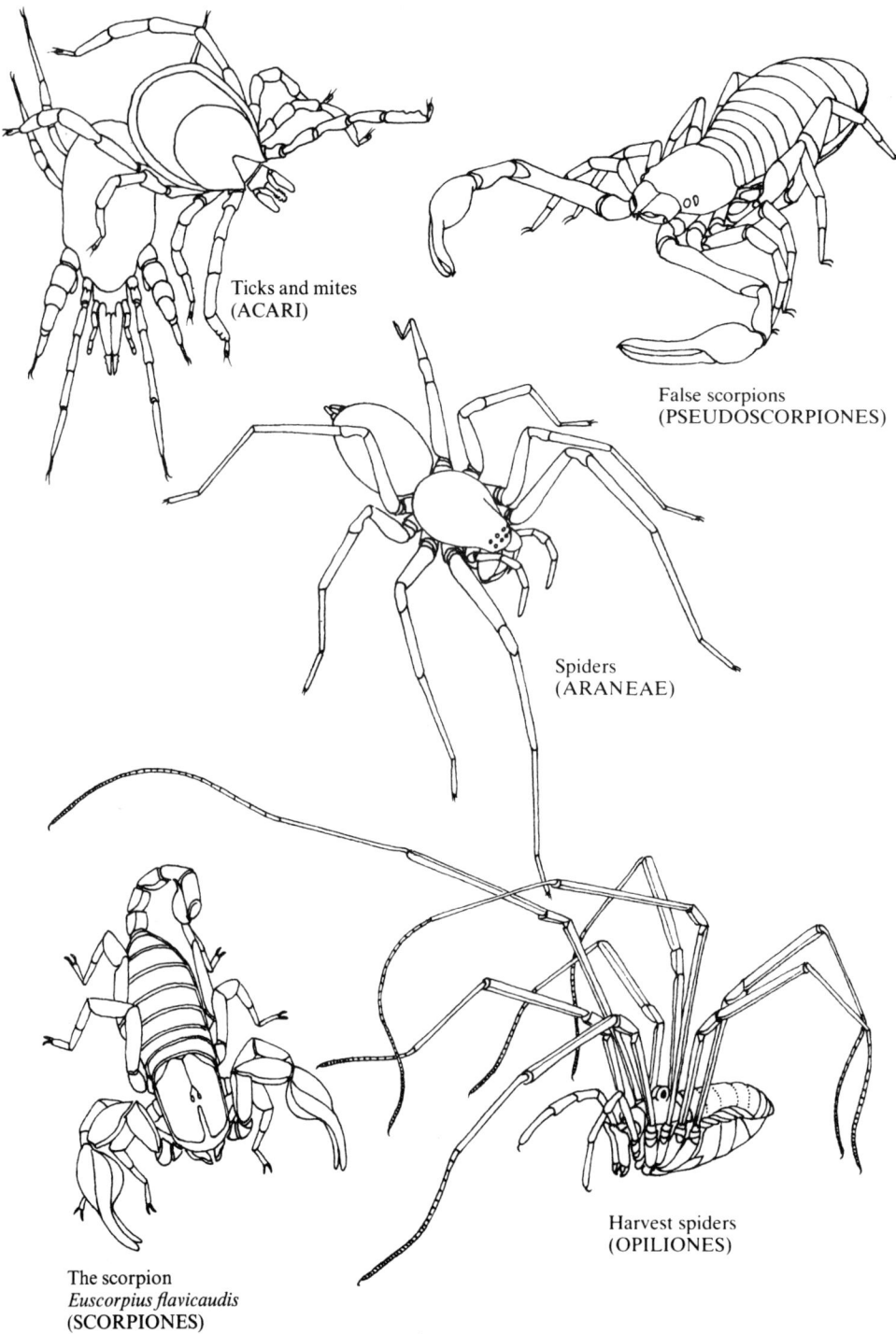

Fig. 1.
A spider and its other British arachnid relatives.

Biology and Ecology

Well over 600 species of spider have been recorded in the British Isles. Almost every conceivable terrestrial habitat has its own characteristic spider community and species. There are no truly marine species—the so-called "sea spiders" belong to the Class Pycnogonida.

Certain species will be seen to come from particular parts of their favoured habitats. Some exhibit striking morphological and behavioural adaptations. For example, a small number of British spiders are exceptionally good ant-mimics. They are often found in close association with ants and, as well as looking like them, they also move and behave like them.

Spiders are an important component of the terrestrial ecosystems in which they occur, and they are often present in great numbers. Their abundance combined with their feeding biology makes them extremely significant in the food chain; they are exclusively carnivorous, generally seizing live prey in the form of other invertebrate animals.

Several spider families use silk to create complex snares in which they trap their prey. Together with the use of venom (supplied from poison sacs in the thorax and secreted through ducts in the fangs), this has earned them a place in the folklore and mythology of nearly every major culture in the world. However, web construction is by no means the only method that they employ to catch their prey. Among the British families there are a large number of species that hunt by stealth or use camouflage to remain undetected until their final strike. Others are ground-running species that seize their prey when on the move, although some of these may also adopt a "sit-and-wait" strategy and allow the prey to come to them.

The diversity of predation strategies is matched by the variety of courtship behaviours employed within the different spider families. In order to pacify the normally aggressive females, males are often required to engage in a bewildering display of leg-waving, web-jerking, vibrating and palp semaphore before their advances are accepted. The males of one species (*Pisaura mirabilis*) even present the female with a small silk-wrapped insect "gift" in order to distract her during the process of mating (Fig. 2).

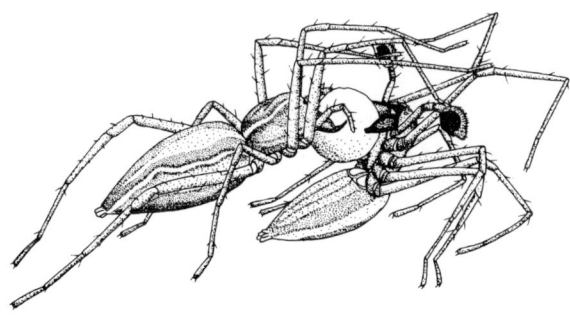

FIG. 2
Part of the mating process of *Pisaura mirabilis*, the nursery web spider. The male is presenting the female with a silk-wrapped insect gift prior to copulation (after Bristowe, 1958).

TABLE 1. *Distinguishing the sexes and maturity of spiders.*

	BODY FORM	**PEDIPALPS**	**ABDOMEN (BELOW)**
MALE		*Modified* to carry a copulatory organ of varied complexity.	*Without* an epigyne in the centre of the abdomen, immediately in front of the epigynal fold (although may be darker in this area).
FEMALE		*Simple* (although females of the linyphiid genus *Agyneta* have swollen palpal tarsus-(p.t.)).	*With* a chitinous plate or process in the centre of the abdomen immediately in front of the epigastric fold—the epigyne (e). Absent in some adult females
IMMATURE	Resemble adults in general appearance but lack modified palps and epigyne. Often very pale or, in some cases, almost translucent, and appear more fragile and less robust than adults.	*Simple* (although immature males have a swollen tarsus (t) during later instars).	*Without* an epigyne in the centre of the abdomen, immediately in front of the epigastric fold. Immature females may be swollen in this area.

Contrary to popular belief, male spiders are only rarely killed and eaten by the females and often live to mate several times with the same or other females. In many instances, the male and female share the same web for considerable periods of time. Table 1 shows the main characteristics for distinguishing male, female and immature spiders. In mature males the final segment of the pedipalps (the leg-like mouthparts at the front of the body) are modified to carry a more-or-less complicated copulatory organ. Immature and female spiders have a simple pedipalp without a modified final segment.

Most adult females are distinguished from immatures by having a sclerotised plate or process (called an epigyne) in the centre of the lower abdomen immediately in front of the epigastric fold. The epigyne is at the entrance to the female reproductive organs and receives the male pedipalp during copulation. Its structure varies in complexity between species and adult females of the families Dysderidae, Oonopidae, Scytodidae, Atypidae and Tetragnathidae have no visible epigyne. There are often considerable differences between the size and shape of males and females of the same species.

Identification to species is usually based on the structure and shape of the palp and epigyne of adult spiders.

As well as having no modified pedipalps or epigyne, immature spiders may be recognisable through a number of other features. They are often very pale or, in some cases, almost translucent and appear more fragile and less robust than adults.

Spiders have a tough outer skeleton called the cuticle, which does not grow with the rest of the body and has to be changed periodically for a larger one. The change is called moulting or ecdysis. When about to moult the spider stops feeding and becomes quiet for several hours or days, often in the safety of a silken cell or retreat, while the inner layers of the skeleton are dissolved away and a soft new cuticle is secreted under the old outer layers. When it is ready, the spider puffs itself up by a combination of swallowing air and muscular action and splits open the old cuticle. It then drags itself clear of the old skin and the new cuticle gradually hardens over its newly-expanded body.

The stages between moults are called instars. The young resemble the adults in general appearance and they often have similar ecological preferences and behaviour. The resemblance increases with each moult and the later instars may be identified to family using the keys and tables in this paper. Most spiders grow to maturity and mate in the course of a single year. However, it is not unusual for certain species to take several years to reach adulthood. Males eventually die after mating, females after laying eggs.

All spiders use silk whether they make webs or not. The presence of the small appendages which produce the silk, the spinnerets, at the hind end of the abdomen is one of the characteristics that distinguishes them from all other invertebrates. Silk is used to construct a retreat or nest which may be a tube, a silk-lined excavation or rolled leaf, or a cell under bark or stones. The water spider forms a bell-shaped dome under which air can be trapped. Silk is used by females to construct egg sacs and by males to make sperm webs. The snares, if they are constructed, vary between funnels, sheets, irregular meshes and the regular geometry of the orb web. Most spiders pay out a "drag line" as they move about and for species which live above the ground this serves as support should they fall or jump. Occasionally, they may stop and attach it to the substrate by a number of looped threads, forming an "attachment disc". The drag lines of ground-running species provide evidence of their abundance when, on dewy mornings, they can often be seen to almost completely clothe the ground.

One of the most interesting uses of silk is as a means of long-range dispersal. While many spiders are extremely active and may be able to move quite considerable distances along the

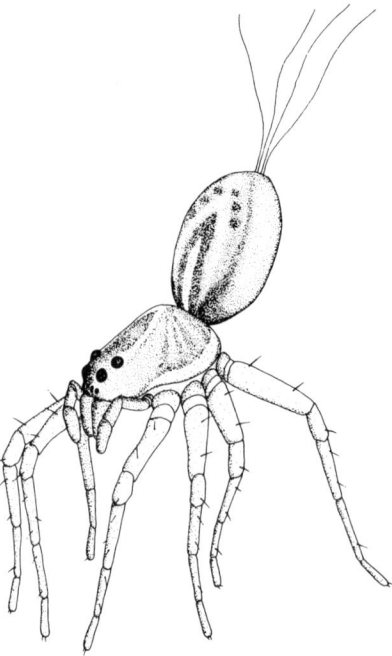

Fig. 3.
A juvenile spider in a characteristic pose, sending out lines of silk into the wind prior to "ballooning"—an aerial dispersal mechanism which allows spiders to travel many hundreds of miles.

ground, they possess no obvious method for making longer journeys. However, they are able to use silk to peform an activity known as "ballooning". By paying out silk while standing at the top of grass stems or posts they are able to float away on the prevailing wind and may travel many hundreds of miles. Figure 3 shows a young spider in characteristic pose, standing on "tip-toes" with its abdomen pointing upwards, sending out lines of silk into the wind.

A number of books provide excellent introductions to the biology and ecology of spiders. The best of these is by Bristowe (1958) who gives a broad background to the ecology and behaviour of all the major families, while Foelix (1982) provides a more detailed consideration of many aspects of spider biology.

Finding and Collecting Spiders

Spiders may be found in almost every kind of habitat and an impressive list of species, from a range of families, can be accumulated by searching in and around the average house and garden. However, it will soon become apparent that the best sites for collecting are those which have been least disturbed by modern agriculture and forestry. While corn fields, rye-grass leys and conifer plantations contain some interesting species, there are many more to be found in flower-rich meadows and grasslands, the margins of streams and ponds, marshland, mountain and ancient broadleaved woodland.

With a certain amount of experience, it is possible to identify many spiders to family level in the field; a number may even be identified to species. In the beginning, however, accurate identification will involve catching them and taking a closer look. This may remain a necessity with many of the smaller species even for experienced collectors. The basic methods for collecting spiders are detailed below and illustrated in Figs. 4 and 5.

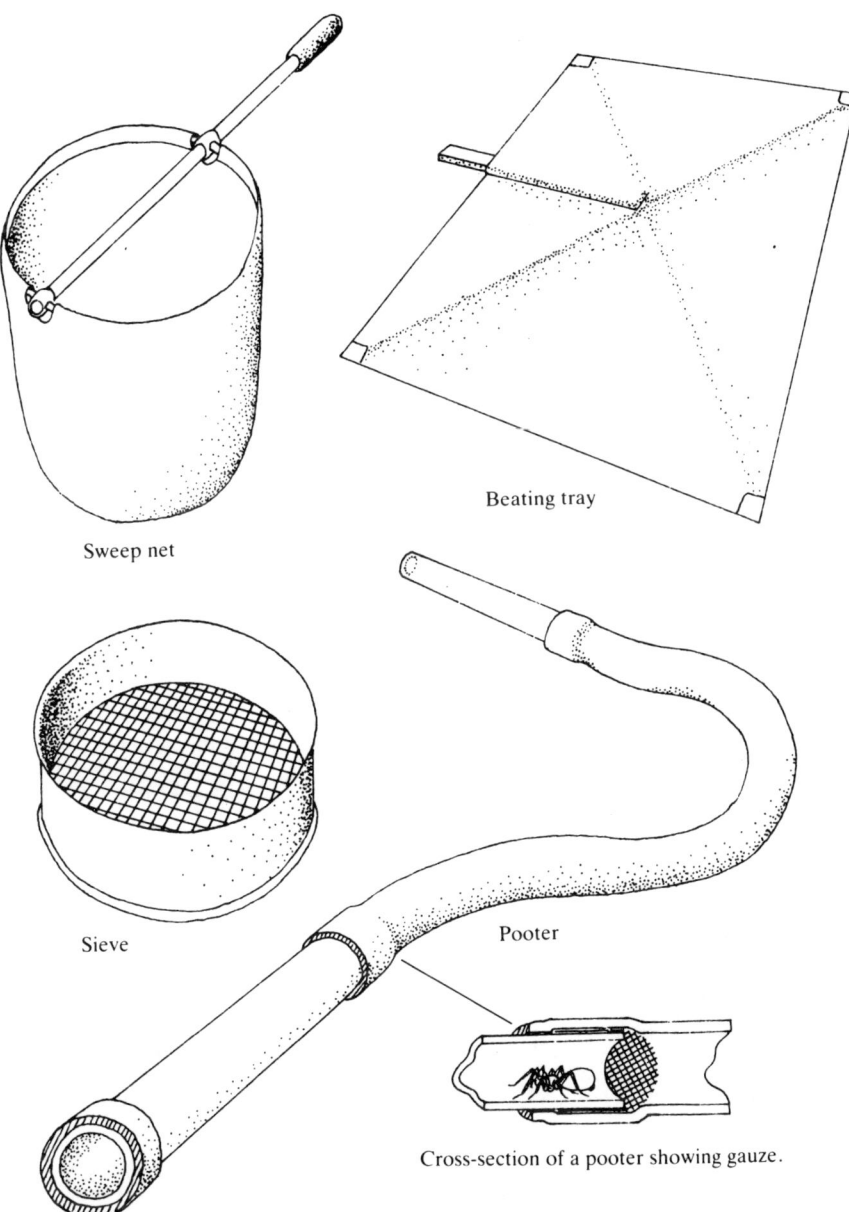

FIG. 4.
Collecting methods.

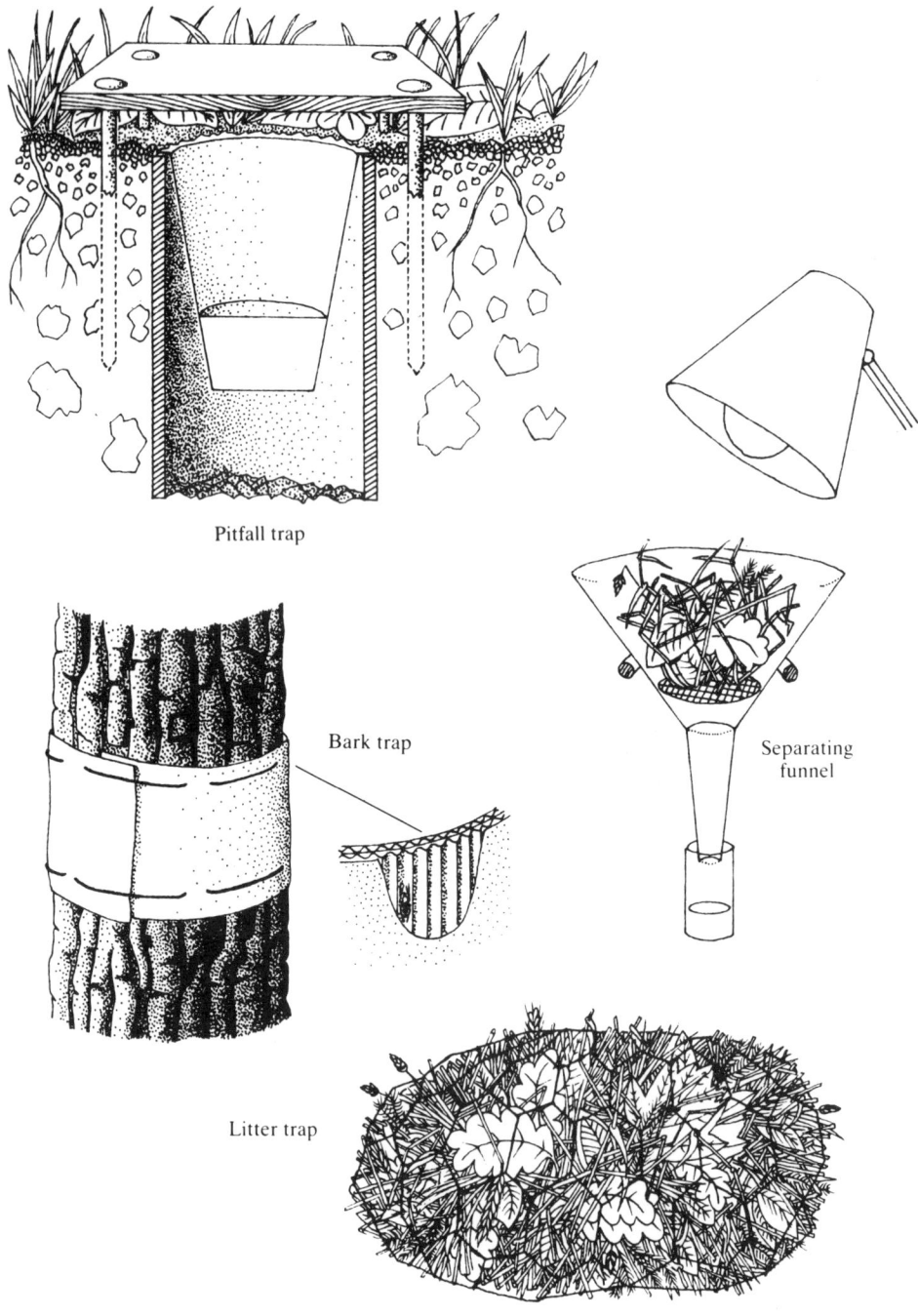

Fig. 5.
Collecting methods (continued).

Sweep Net. The spiders which spend their time in rough grassland and low herbage, either in webs or walking on the vegetation, can be collected using a robust short-handled net with a sturdy canvas collecting bag, reinforced around the rim. The net is swept to-and-fro through the vegetation as the operator walks forwards, stopping at regular intervals to examine the contents and remove any spiders gathered. A butterfly net is unsuitable for this method of collecting as it quickly snags or tears on thorns and briars.

Beating Tray. Spiders living in the foliage of trees and bushes, out of reach of the sweep net, can easily be collected using a beating tray and a walking stick. The tray is held under the vegetation to collect any spiders dislodged when a branch or foliage is hit sharply with the stick. The tray consists of a sheet of some durable material stretched over a collapsible wooden or steel-tubing frame about 1 metre square. Alternatively, a smaller, solid plastic tray with a 5–10 cm lip may be used. This contains the spiders, which often rush about madly when dislodged, allowing the operator time to pick them up. The plastic tray can also be used when beating low vegetation including grass tussocks, rushes or reeds, or the overhanging herbage of ditches, small streams and earth banks.

Pitfall Trap. It is often difficult to catch ground-running or night-active spiders using the methods detailed above but there is another simple technique for catching these species. Pitfall traps may be set out using jam jars, plastic cups or other suitable receptacles dug into the soil, flush with ground level. Wandering spiders and other invertebrates fall in (often in great numbers during the summer months) and are unable to escape.

In order to stop the receptacle filling with water and leaf litter and to deter some of the larger predators (shrews and mice) the trap may be covered with a flat stone supported by three or four smaller stones or a square of plywood with a six-inch nail through two corners. If the trap is left dry it will have to be visited regularly to stop the animals that fall in from eating one another. Alternatively, a small amount of ethylene glycol (anti-freeze) may be poured in to a depth of three to five centimetres. The ethylene glycol is an alcohol and will kill and preserve the specimens that fall in.

Bark Trap. This is a simple technique which is extremely effective in sampling the fauna of tree bark. Two 20–25 cm wide strips of corrugated brown paper, their length determined by the diameter of the tree, are tacked together with a darning needle and tied around the trunk. The corrugations provide refuges for many small invertebrates and after several weeks the trap may be removed, placed in a closed bag and later dismantled over a tray or large sheet by slowly teasing apart the corrugations. A modification to this method involves using strips of bubbled polythene wrapping covered with a strip of black polythene to exclude light. This has several advantages: it is reusable, weatherproof and it is possible to see the trapped spiders through the polythene, speeding up the sorting process. In both cases efficiency can be improved by using twigs or shavings to create additional cavities between the two strips.

Litter Trap. An alternative to sieving litter in the field or to collecting leaf litter or vegetation and sorting it in the lab, is to construct litter traps. These consist of straw, wood shavings or naturally occuring vegetation litter in a closed chicken-wire basket. The trap is placed in the desired habitat for at least a week allowing spiders and other invertebrates to take up residence. An interesting variation is to tie the litter trap high in the boughs of a large tree where it will attract spiders normally associated with high canopy and birds' nests and not often collected. Once recovered, the traps may be sorted in the lab or at home, the trap replenished and put out again. This method, which can be used at any time of the year, is a valuable technique for sites where the vegetation is not suitable for the more robust collecting methods.

Vegetation removal. This traditional method involves the removal by sawing-off, or tearing-off, whole grass tussocks, clumps of moss and other vegetation. These may then be sorted on a tray in the field and any spiders removed, or taken back to the lab in a plastic bag to be sifted through at leisure. Nowadays, with habitats under increasing threat from a variety of pressures, these techniques are becoming less and less desirable. They can cause considerable damage to the vegetation and should only be carried out if there is likely to be minimal impact on the area in question (and if the necessary permission has been obtained from the owner, or warden if the site is a nature reserve).

Sieving. Many spiders are found in leaf litter but can be difficult to extract. If a simple garden sieve or a chip-frying basket is taken into the field the litter may be placed in it and shaken over a tray. By collecting litter in plastic bags a similar operation may be carried out in the relative comfort of the lab or at home.

Separating Funnel. Extracting spiders from litter, moss, dead wood, other vegetation and even old bird and mammal nests can be difficult and time consuming. The separating funnel is a passive method for accomplishing this task. A mesh screen is placed in an opaque funnel, supported above a collecting tube. The sample is placed on the mesh and a bright light, such as a desk lamp, positioned over the apparatus. A combination of heat and light encourages the inhabitants of the sample to move downwards and eventually fall through the mesh into the collecting tube. The tube may be filled with preservative but live collecting is possible using a dish lined with damp blotting or filter paper, which must be examined regularly. A week is usually long enough to extract nearly all of the occupants from a 10 cm^3 litter sample.

Hand Collecting. This method, as the name suggests, is the most basic of the techniques and requires little in the way of equipment—apart from water-proof clothing if the ground is wet. It simply involves searching the vegetation, often by crawling along on all fours, separating the grass, leaves and litter by hand. It is a selective method, the collector needs only to pick the specimens he or she wants, an advantage it holds over certain of the techniques described above.

The Pooter. This elegantly-named piece of equipment is probably the most essential tool in the spider-collector's kit. It is a technique for picking up small to medium-sized spiders without the need to use forceps or fingers and it, therefore, avoids possible damage to the specimen. The spider is sucked through an arrangement of rubber and glass tubing with a gauze filter to prevent it arriving in the collector's mouth! There are two basic designs: the "suck-blow" pooter is used to pick up the specimen and then deposit it in a separate collecting tube while the "composite" pooter contains an integral tube that may be removed, sealed and replaced with a fresh tube.

Collecting tubes and jars. These come in a range of shapes and sizes. When in the field it is often desirable to carry a variety of glass tubes stoppered with plastic or cork as well as some plastic screw-top containers to hold larger, live specimens. Live spiders kept in the same tube invariably eat or damage each other before they reach the laboratory and should, therefore, be kept apart. For many of the tiny spiders and for some field projects, it may be necessary to kill the animals by placing them in 70% alcohol; the reasons for this are given in the sections on examining and preserving spiders and, in particular, spider conservation. It may be useful to carry a supply of alcohol in the field.

Ecological notes. All species have a definite set of characteristics which, when taken together, help to determine their identity. These include aspects of their ecology and behaviour as well as of their physical appearance. When using the keys in this paper, some knowledge of these ecological characteristics will make the task of identification to family

Keys to the Families of British Spiders

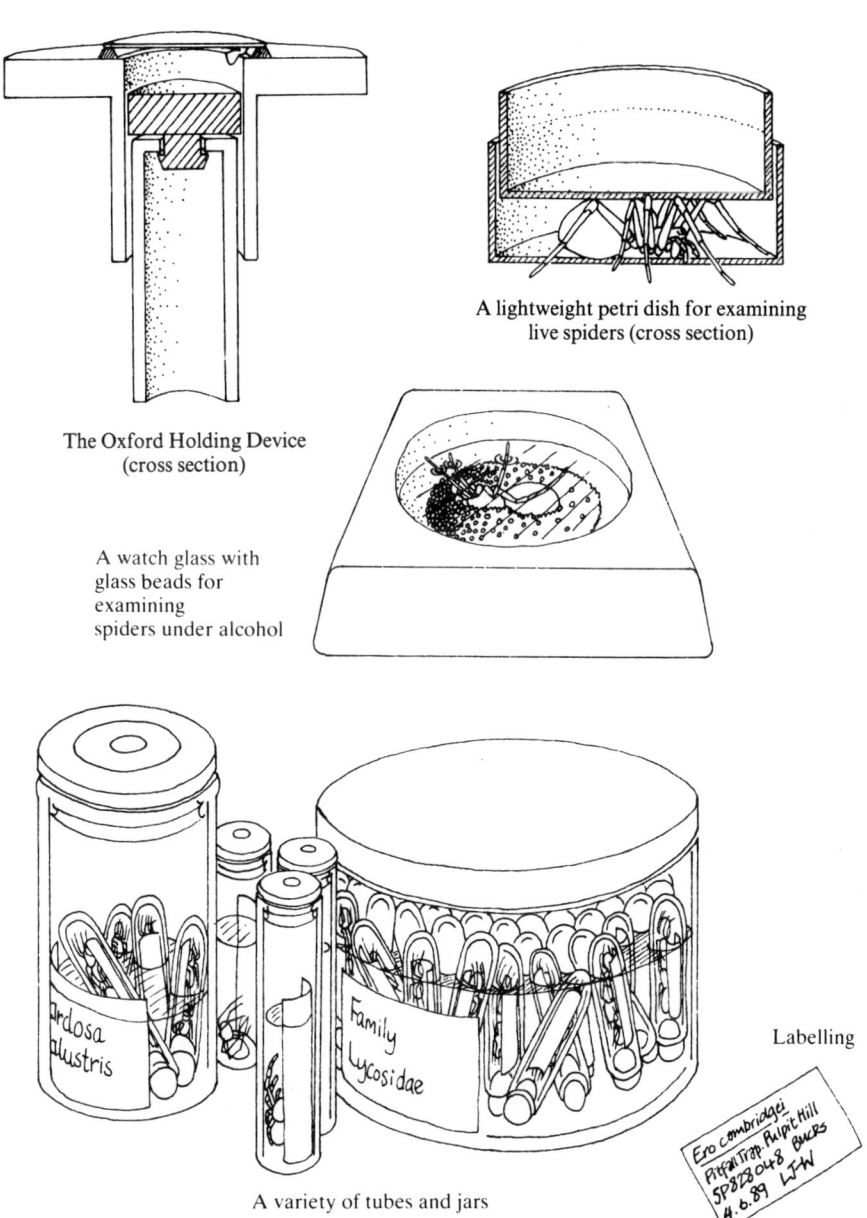

FIG. 6.
Examining spiders.

level considerably easier. When collecting specimens in the field, it is extremely valuable to record their habitat and ecological behaviour. For example: "collected in a large circular web at the top of grass stems" or "wandering on a tree trunk, no web". These notes can be numbered to correspond with a marked tube or can be torn off and placed in the tube with the specimen. Use a pencil; most inks dissolve in water or alcohol.

Finally, every collector needs a sturdy canvas bag to carry all this paraphernalia! Chalmers and Parker (1989) provide very useful further reading on practical fieldwork techniques while Southwood (1979) is a slightly more detailed and technical text.

Examining Spiders

With experience, some specimens may be identified to family level with the naked eye or with a low-powered ($\times 10$) hand lens. However, a binocular microscope with a magnification between $\times 20$ and $\times 80$ will be necessary for the beginner. Regardless of expertise, identifying the smallest spiders to family (and the majority of spiders to species) will invariably require a microscope. In order to observe detailed structures under the microscope it is also important to have a source of bright illumination.

Identifying Live Spiders. Identification can be attempted with live specimens. This is particularly desirable when they are not required for a reference collection and may be returned to the population from which they came, or can be kept alive for study in the laboratory. Large, robust specimens can be held in place by trapping the animal between the lid of a small plastic petri dish and the inverted base (see Figure 6). Small dishes, 5 cm in diameter, are particularly useful. These are easily manipulated so that both sides of the specimen can be examined, although care should be taken to avoid crushing the spider.

A slightly more elegant piece of apparatus, suitable for more fragile small to medium-sized spiders, has been developed (Oxford, 1981) and is illustrated in Figure 6. It may be constructed in minutes out of cheap, readily-available materials. Oxford (1981) describes the construction and operation as follows: "... It consists of the barrel of a 5 ml plastic disposable syringe cut off about 1.5 cm from the non-pointed end. The plunger is cut off about 2 cm from the base of the rubber piston, which must be the flat-topped type. A shallow notch is filed across the top of one side of the barrel between the finger lugs to allow the entry and exit of air during use. A 1.6 cm diameter cover slip is glued across the end of the barrel with epoxy resin, and the plunger inserted into the barrel so that the flat rubber piston faces the coverslip. The ease of movement of the piston should be adjusted by uniformly abrading the sides of the piston with emery paper until the piston can be moved easily, without jerks, up and down the barrel. Finally, a support for the apparatus is constructed by taking a circle of perspex, wood or card and cutting a hole through the middle of a diameter slightly larger than the diameter of the syringe barrel. This support is used in place of the glass disc on the stage of a binocular microscope. In use the apparatus sits in the hole supported by the finger lugs of the barrel.

The holding device is used as follows. The plunger is removed; a spider is introduced into the barrel and the plunger replaced. If the barrel is held 'coverslip up' while the plunger is moved, the spider will orientate so that the dorsal surface can be examined. If the ventral surface is of interest, the barrel is held the other way up during the final adjustment of the plunger. The plunger is moved using the thumb and index finger in a ratchet-like movement against the side of the barrel until the spider is lightly held between piston and coverslip."

Identifying Preserved Spiders. It is often necessary to kill spiders for the purposes of scientific research and to establish reference collections. The cuticle of spiders is not as hard as that of insects and, once killed, will quickly dry out and shrivel-up if left in air. For this reason they are best studied completely immersed in 70% alcohol, in a dish or watch glass. The specimen will have to be viewed from various angles so it is best to place a layer of washed sand or fine glass beads in the bottom of the viewing receptacle to hold the specimen in the desired position without damaging it (Fig. 6). It is also important to have a pair of fine forceps, and a seeker, to pick specimens out of tubes and to manipulate them under the microscope.

Spiders that have been killed and stored in alcohol are easier to examine than live specimens because they are more readily moved into the desired position. This is particularly important when identifying to species, when a range of different characters may need to be viewed one after another, or frequent reference made backwards and forwards between characters. It is also easier to see certain important characters under alcohol.

Identifying Spiders to Species. There are a number of keys available to identify spiders beyond the family level. Locket and Millidge (1951 and 1953) key out most of the British species and are supplemented and updated by a third volume (Locket, Millidge and Merrett, 1974). Most up-to-date is the comprehensive three-volume guide by Roberts (1985 and 1987). Jones (1983) provides a very useful field guide with colour photographs of over 350 representative species.

Keeping and Preserving Spiders

Once identified, live specimens may be released in the habitat from which they were originally collected. Alternatively, a great deal may be learned about feeding, mating and web-spinning behaviour by keeping spiders in captivity. Murphy (1980) shows how to keep a range of invertebrates, including several of our native spiders.

Spiders killed and preserved in alcohol can represent an extremely important scientific resource. They can provide a reference for other collectors with which to compare and contrast their own specimens—a valuable aid to identification. Of equal significance is the information collected with them relating to their habitat, behaviour and location. When collated this can give clues to their distribution, rarity and conservation. With difficult specimens or infrequently seen rarities, it may be necessary to send them off to have identifications confirmed by a "spider expert".

Once identified, spiders should be kept in plastic-stoppered glass tubes filled with 70% alcohol and labelled with the collecting data. Larger collections may be kept in small tubes stoppered with cotton wool and inverted in alcohol filled jars (Fig. 6). Spiders can lose their colour, or even bleach, in direct sunlight so specimens should be kept in the dark, preferably away from strong heat.

Spider Conservation

It is obviously better to identify spiders alive and then return them to where they were collected. While this will satisfy the unease that many feel about having to kill invertebrates in order to study them, there is still a strong case for building up a reference collection.

In order to foster the conservation of any species, it is important to learn more about its ecological requirements. By collating distribution records, a picture of the occurrence of individual species may be pieced together. From the resulting distribution maps, it is possible to identify rare species and, if the records are accompanied by reasonably detailed field notes, to specify their ecological preferences. All species rely on a particular set of environmental conditions for their survival and, once these are identified, appropriate measures may be taken to encourage their conservation by managing a habitat to maintain or create these conditions.

Detailed studies of the spider communities associated with different habitat types can reveal the existence of assemblages of species. Some of these assemblages may themselves be rare and, within them, it may be possible to identify species that suggest certain attributes about the environment from which they came. These can relate to the animal or plant community and may be used as indicators of "quality" or conservation value.

A spider recording scheme has been set up jointly by the British Arachnological Society and the Biological Records Centre (which is part of the government-funded Institute of Terrestrial Ecology). Further details of this scheme may be obtained from: The Biological Records Centre, Institute of Terrestrial Ecology, Monks Wood Experimental Station, Huntingdon, Cambridge PE17 2LS.

Collecting Code

It is necessary to keep a reference collection in order to carry out serious recording or community studies, particularly if there is a need to obtain confirmation from experts for species which are difficult to identify. It is unlikely that collecting alone has caused the extinction of any spider species in the British Isles. However, with the increasing loss of habitats resulting from forestry, agriculture, industrial, urban and recreational development, the point has been reached where a Code for Spider Collecting is required in the interests of spider conservation. Insect conservationists have already issued a "Code for Insect Collecting" (Joint Committee for the Conservation of British Insects) on which the following is based:

1. General Collecting
   - Readily identified spiders should be examined alive and released where they were captured.
   - No more specimens of any species than are strictly required for any purpose should be killed.
   - Do not take a species year after year from the same locality.
   - Try to visit new sites rather than collect a local or rare species from a well-known or over-worked locality.
   - Specimens for exchange, or disposal to other collectors, should be taken sparingly or not at all.
   - If a trap used for scientific purposes is found to be catching large numbers of local or rare species it should, if possible, be re-sited. Live-trapping is to be preferred.
   - Consideration should be given to photography instead of collecting, particularly in the case of large "showy" species.
2. Permission and Conditions for Collecting
   - Always seek permission from a landowner (or occupier) when collecting on private land and always comply with any conditions that may be stipulated.
   - When collecting on nature reserves, or sites of known conservation interest, supply a list of species collected to the appropriate authority and observe the codes listed below.
3. Protecting the Environment
   - Do as little damage to the environment as possible. Be careful of disturbing nesting birds and trampling vegetation, particularly rare plants.
   - Only remove vegetation and leaf litter if you have specific permission. Replace moss (and any other form of vegetation that is likely to recover) in its appropriate habitat once it has been worked for spiders.

- Use suitable restraint with any collecting method that involves disturbance to the vegetation. (For instance, when "beating" trees and shrubs never thrash the vegetation so hard that large quantities of foliage, twigs and branches are removed.)
- Replace logs and stones after searching beneath them, and replace the bark from dead timber.
- Adhere to the "Countryside Access Charter" (Countryside Commission, 1985).

Two British spider species, the fen raft spider (*Dolomedes plantarius*) and the ladybird spider (*Eresus niger*) are protected by law and may only be collected under licence.

The British Arachnological Society

For anybody seriously interested in spiders, or other arachnids, it is well-worth joining the British Arachnological Society which publishes a bulletin and newsletter. These are produced three times a year and contain information on collecting methods, new distribution records, and notes on the ecology and taxonomy of arachnids.

The Society has a number of regional groups who run collecting days and weekends as part of the spider recording scheme. These are often combined with identification "workshops" and are an excellent way of meeting other members and improving identification and field-craft techniques. The Field Studies Council runs residential field courses on spiders which are usually led by experienced members of the British Arachnological Society.

Further details may be obtained from the Society's Membership Treasurer: S. H. Hexter, 71 Havant Road, Walthamstow, London E17 3JE.

Acknowledgements

Thank you very much to all those people who offered written or verbal comments in response to the Test Version of the key, many of which have been incorporated and all of which have hopefully increased the value of this as a working document. Thanks also to Mr G. H. Locket for his encouragement throughout and to Mrs Frances Murphy. To Dr Stephen Tilling whose knowledge of the mechanisms of key construction has been invaluable and to Mr Ivor Lansbury and the Hope Entomological Collections, and Mr Peter Harvey for the loan of specimens. Finally, many thanks to my wife Clare and daughters Hannah and Miriam for their support and the sacrifices they have made in order to see this key completed.

External Features and Glossary of Terms

This glossary, and the accompanying figures, form a reference while using the keys. Terms highlighted by italics are also referred to in other parts of the glossary.

ABDOMEN: (Fig. 7a). The second, or *posterior*, of the two major body divisions (sometimes referred to as the *opisthosoma*).

ANTERIOR. Towards the front.

ANTERIOR LATERAL EYES: (Figs. 9a & 9b). The outer pair in the frontal row of *eyes*.

ANTERIOR MEDIAN EYES: (Figs. 9a & 9b). The inner pair of the frontal row of *eyes*. Six-eyed spiders lack these.

BALLOONING: (Fig. 2). The process of aerial dispersal by spiders, using silk.

BOOK LUNGS: (Fig. 7b). Respiratory organs situated on the *ventral* surface of the *abdomen* with the *lung slits* opening to the outside. Usually a pair but one family, the Atypidae, has two pairs.

BOSS: (Fig. 7d). A well-developed shiny prominence at the base of each *chelicera*. May or may not be present.

BRISTLE: (Fig. 8a). An obvious feature. More robust, but may be longer or shorter than a *hair*. Less, robust than a *spine*.

CALAMISTRUM: (Fig. 8g). A series of curved *bristles* running along the *dorsal* edge of *metatarsus IV*. Present in some families only. It is used to comb out a viscid substance extruded by the *cribellum* which, when combined with ordinary threads, provides a composite and adhesive lace-like strand of faintly bluish colour.

CARAPACE: The structure covering the upper surface of the *cephalothorax*.

CARDIAC AREA: (Fig. 7a). An area, often marked by a special pattern, that extends back along the mid-line from the front of the upper surface of the *abdomen*. Positioned over the heart, hence the name, and only notable in some species.

CEPHALOTHORAX: (Fig. 7a). The anterior of the two major body divisions. (Sometimes referred to as the *prosoma*).

CERVICAL GROOVE: (Fig. 7a). A furrow on the *carapace* which extends forward and to the sides, marking the boundary between the *head region* and the *thoracic region*. It is sometimes indistinct or completely lacking.

CHELICERAE: (Figs. 7b–7e). The jaws (singular Chelicera), thickened and robust at the base, narrowing towards the outward tip, with a *fang* articulated at the outward corner.

CLAWS: (Figs. 8a–8f). Terminal appendages on the *legs* and *pedipalps*. All families have at least two per leg but some also have a median, unpaired, third claw. This is often small and difficult to see because of the *bristles* located in this region.

CLAW TUFTS: (Fig. 8f). Many (although not all) two-clawed spiders have a dense tuft of hairs at the tip of the leg *tarsus*.

CLYPEUS: (Fig. 9a). The area between the anterior row of *eyes* and the front edge of the *carapace*.

COXA: (Fig. 7a). The joint of the *legs* and *pedipalps* which is nearest to the body.

CRIBELLUM: (Fig. 8j). A plate, covered with minute *spigots*, situated in front of the anterior *spinnerets*. It produces a viscid substance which is combed-out by the *calamistrum* and combined with ordinary threads from the *spinnerets* to provide a composite, adhesive lace-like strand of a faintly bluish colour.

CUTICLE. The hardened layer which covers the body of a spider. It combines rigidity with flexibility. In addition to its protective function, the cuticle determines the form of the spider; its relative impermeability to water reduces desiccation and it provides a firm base for the attachment of muscles. It remains soft at the joints to allow movement.

DIMORPHIC. Occurring in two distinct forms of the same species.

DISTAL. Terminal. At the outer end of (a *leg* or *pedipalp*).

DORSAL. The top or upper surface.

EPIGASTRIC FOLD: (Fig. 7b). The fold or furrow across the *ventral* surface of the *abdomen* that usually joins the *lung slits*.

EPIGYNE: (Fig. 7b). The *sclerotised* plate, located in front of the *epigastric fold* forming part of the adult female's reproductive organs. May be difficult to see and varies in complexity between species.

EYE AREA. The space enclosed by the *eyes*.

EYES: (Fig. 9). Spiders have simple eyes (contrasting with the complex compound eyes of the insects). Most spiders have eight (four pairs) but some families have fewer. In most families, eyes are arranged in two rows, termed the *anterior* (frontal) and *posterior* (hindmost) rows, with the inner pair in each being termed *median* and the outer pair *lateral*.

FANG: (Fig. 7c). The tooth-like structure which articulates at the outer corner of *chelicera*. Used by the spider to pierce the prey, bearing a small duct through which poison is injected.

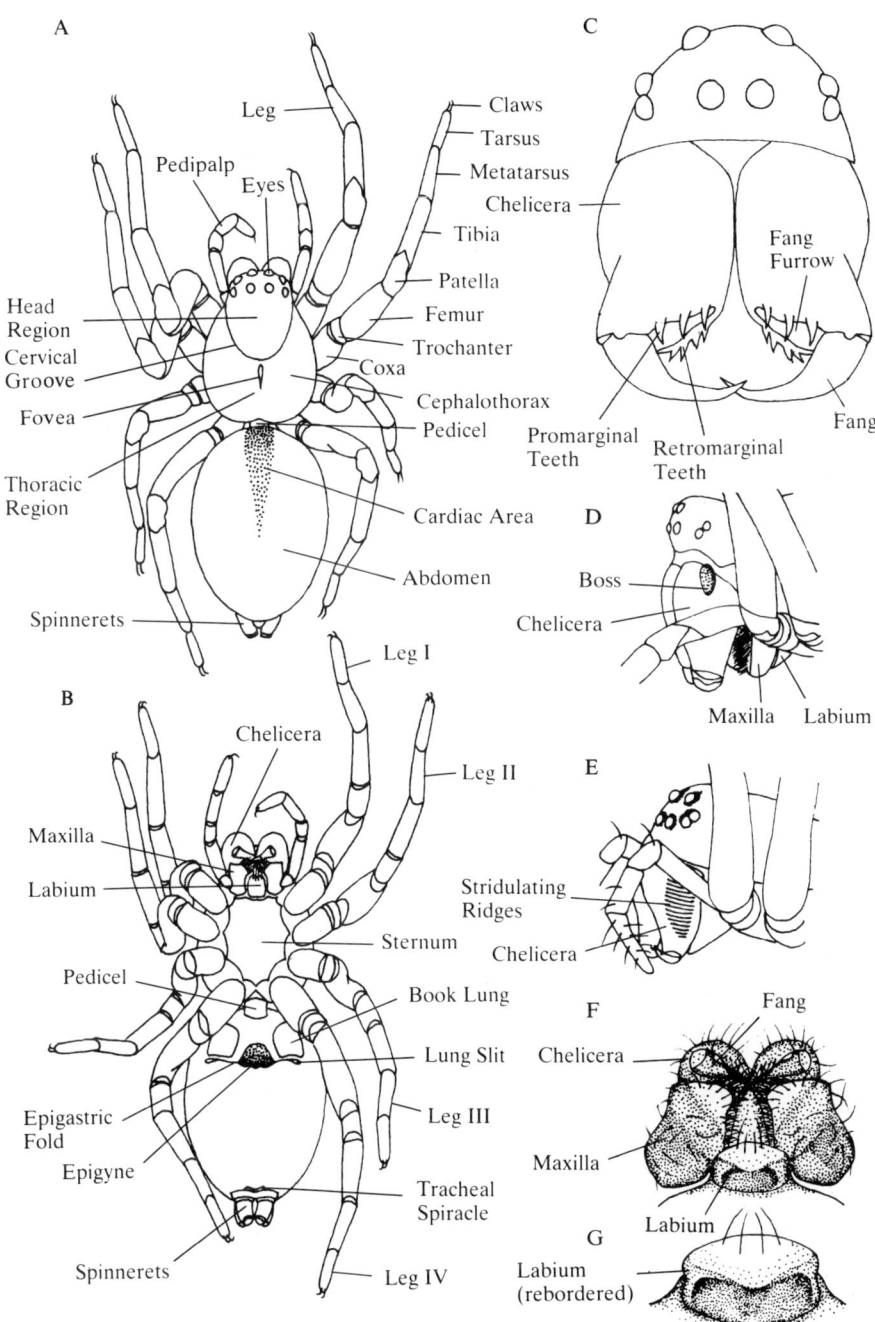

FIG. 7.

Spider: external features.
a) view from above; b) view from below; c) chelicerae and fangs; d) chelicera with boss; e) chelicera with stridulating edges; f) mouthparts; g) rebordered labium.

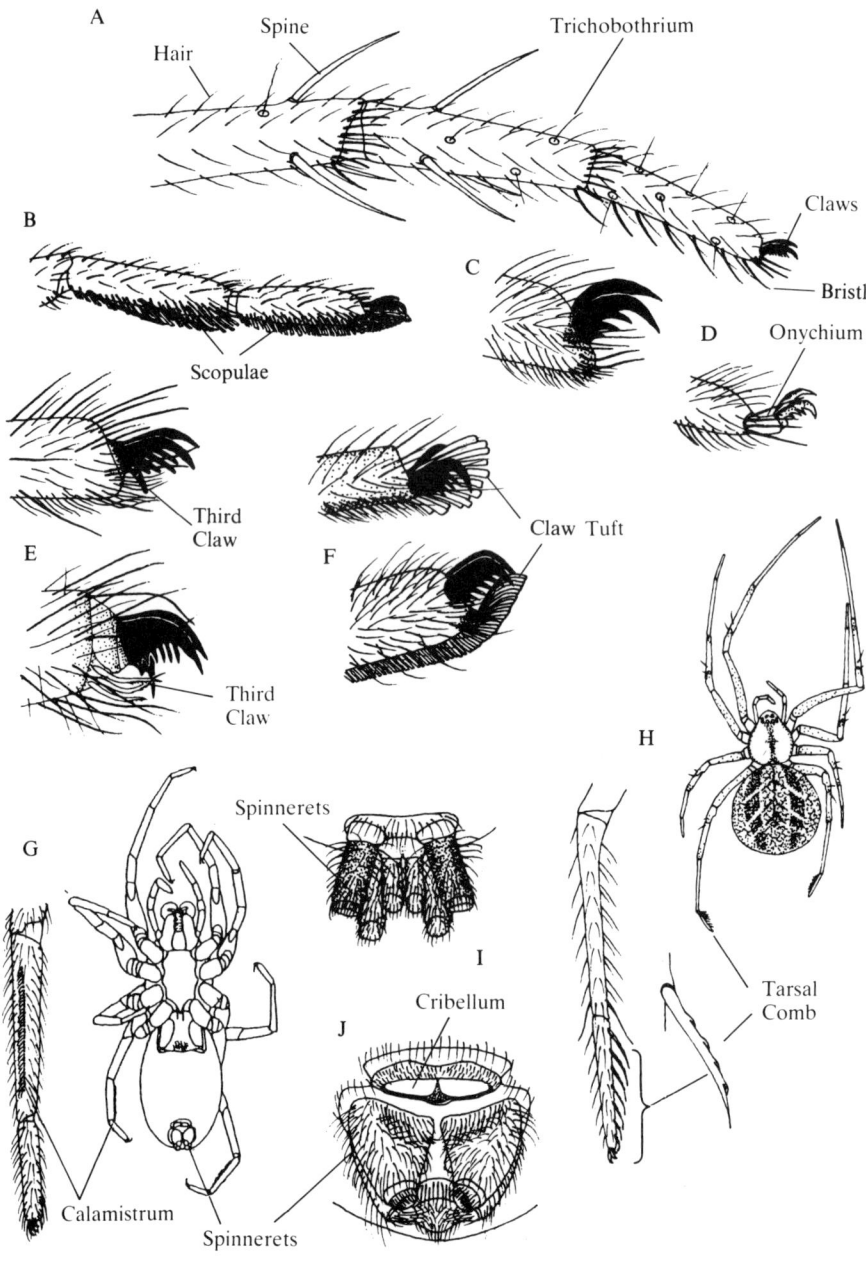

FIG. 8.

Spider: external features.
a) Leg showing spines, bristles, hairs and trichobothria; b) metatarsus and tarsus with scopulae; c) two claws, no claw tufts; d) claws with onychium; e) three claws; f) two claws with claw tufts; g) calamistrum on metatarsus IV; h) tarsal comb on tarsus IV; i) spinnerets, all three pairs visible; j) spinnerets with cribellum.

FANG FURROW: (Fig. 7c). The groove in which the *fang* lies when not in use. Often armed with numerous teeth.

FEMUR: (Fig. 7a). The third innermost segment of the *legs* and *pedipalps*.

FOLIUM. The pattern that sometimes covers the whole *abdomen*.

FOVEA: (Fig. 7a). A conspicuous depression, furrow or pigmented line in the middle of the *carapace*. Not visible in all species.

HAIR: (Fig. 8a). A delicate *setal* structure (see *bristles, spines* and *trichobothria*). Unmodified, arising from the *cuticle*.

HEAD REGION: (Fig. 7a). The *anterior* of the two regions of the *carapace* (Cephalothorax) bearing the *eyes*.

HETEROGENEOUS EYES. Some eyes obviously differ in size, shape or colour from the others.

HOMOGENEOUS EYES. Eyes all of the same size, shape (approximately) and colour.

HYPEREXTENSION. Extension (of the apex of the *metatarsus* in the family Eusparassidae) exceeding that of normal joints. An adaptation for prey capture.

LABIUM: (Figs. 7b, 7f & 7g). The structure which covers the *anterior* part of the *cephalothorax* on the under-surface of the body. It is one of the mouthparts, being located between the two *maxillae*. It is usually free from the *sternum* but is occasionally fused to it.

LATERAL. Towards the sides.

LEG: (Fig. 7a). The organ of support and locomotion. In spiders, they are divided into seven segments: *coxa, trochanter, femur, patella, tibia, metatarsus* and *tarsus*. Each pair of legs is numbered I–IV (from the front to the rear of the body). The *metatarsus* on the second pair of legs from the front would therefore be referred to as "metatarsus II".

LUNG SLITS: (Fig. 7b). A pair of slits immediately below the *book lungs* leading to the lung chambers.

MAXILLAE: (Fig. 7f). The expanded basal segments of the *pedipalps* (singular maxilla). Only visible from below.

MEDIAN. Towards the middle.

MEDIAN EYES: (Figs. 9a & 9b). The one or two pairs of *eyes* in the middle of the *head region*.

MEDIAN OCULAR AREA: (Fig. 9a). The area included by the four *median eyes*.

METATARSUS: (Fig. 7a). The second outermost segment of the *leg*.

NOCTURNAL. Active during darkness.

ONYCHIUM: (Fig. 8d). A flexible extension of the cuticle between the claws and the end of the tarsus found in the families Oonopidae and Scytodidae.

OPISTHOSOMA. See *abdomen*.

PALPS. See *pedipalps*.

PATELLA: (Fig. 7a). The fourth innermost segment of the *legs* and *pedipalps*.

PEDICEL: (Figs. 7a & 7b). The waist-like stalk connecting the *cephalothorax* to the *abdomen*.

PEDIPALPS: (Fig. 7a). *Leg*-like mouthparts, but consisting of only six segments (compared with seven in the legs—they lack the *metatarsus*). Should not be confused with antennae, which are absent in all arachnids. The basal segments are expanded and modified to form the *maxillae*.

POSTERIOR. Towards the rear hind end (of the spider).

POSTERIOR LATERAL EYES: (Figs. 9a & 9b). The outer pair in the hindmost row of *eyes*.

POSTERIOR MEDIAN EYES: (Figs. 9a & 9b). The inner pair in the hindmost row of *eyes*.

PROCURVED EYES: (Fig. 9b). The *lateral* pair of *eyes* in each row further forward (toward the front of the head) than the *median* pair.

PROMARGINAL TEETH: (Fig. 7c). The teeth lying along the forward-facing edge of the *fang furrow*.

PROSOMA. See *cephalothorax*.

PSEUDOSEGMENTS. Secondary segments, forming sub-divisions of the major segments.

RADIAL FURROWS. Three pairs of less conspicuous furrows, grooves or pigmented lines running from the foveal area (see *fovea*) to the edge of the *cephalothorax*.

REBORDERED LABIUM: (Fig. 7g). In some spiders the end of the *labium* is thickened and strengthened (often appearing paler than the rest of the structure). This is described as a rebordered labium. Where the labium is not rebordered the structure appears uniformly and smoothly flattened throughout.

RECURVED EYES: (Fig. 9c). The *lateral* pair of *eyes* in each row further back (towards the back of the *head region*) than the *median pair*. When they are markedly recurved the eyes may appear in three rows.

RETROMARGINAL TEETH: (Fig. 7c). The teeth lying on the backwards-facing edge of the *fang furrow*.

SCLEROTISED (SCLEROTISATION). Hardening of the cuticle. Heavily sclerotised areas are particularly shiny and are often heavily pigmented and dark in colour.

SCOPULA: (Fig. 8b). A dense brush of stiff short *hairs* found on the *ventral* surface of the *tarsi* in some groups of spiders (and also on the *metatarsi* in others).

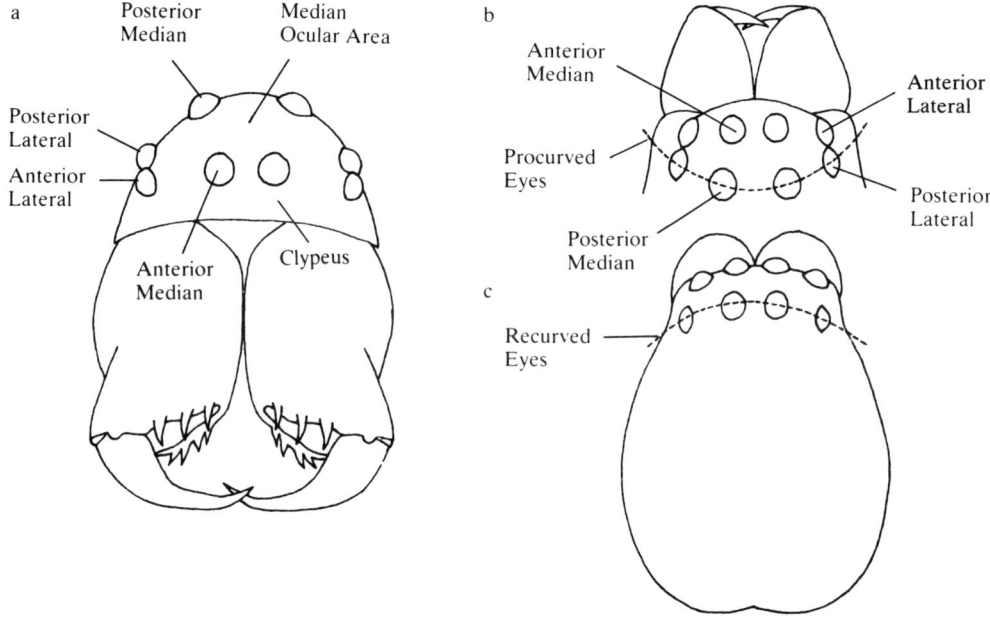

Fig. 9.
Spider external features.
a) Eyes from in front; b) eyes from above, posterior row procurved; c) eyes from above, posterior row recurved.

SCUTUM. A hardened (*sclerotised*) plate covering some or all of the *dorsal* surface and part of the *ventral* surface of the *abdomen*. Only found in a few species of spider.

SETAL (STRUCTURES): (Fig. 8a). Refers to all the spine-like structures (*bristles, hairs, spines* and *trichobothria*) which arise from the *cuticle*.

SPIGOT. A minute peg-like structure. Found in large numbers on the *cribellum* of certain species.

SPINE: (Fig. 8a). The most robust of the *setal* structures. Thicker than a *bristle* and usually longer.

SPINNERETS: (Figs. 7a,b; 8i & 8j). Structures through which silk is produced. Arranged in three pairs situated at the hind end of the *abdomen*.

STABILIMENTUM. Part of some orb webs. A broad band of matted silk threads lying across the centre of the web, the purpose of which is unclear but which may be a device to deflect avian predators.

STERNUM: (Fig. 7b). The posterior of the two major structures covering the *ventral* surface of the *cephalothorax*.

STRIDULATION (STRIDULATING RIDGES): (Fig. 7e). The process of sound production by rubbing together of certain modified surfaces of the cuticle. A number of families, notably the Theridiidae and Linyphiidae, possess stridulating organs which they use to produce high frequency sound and vibration during courtship.

TARSAL COMB: (Fig. 8h). A row of serrated *bristles* along the *ventral* edge of *metatarsus IV*. Found in two families only, the Theridiidae and the Nesticidae.

TARSUS: (Fig. 7a). The outermost segment of the *legs* and *pedipalps*, usually bearing two or three *claws*.

THORACIC REGION: (Fig. 7a). The posterior of the two regions of the *carapace*.

TIBIA: (Fig. 7a). The fifth innermost segment of the *legs* and *pedipalps*.

TRACHEAL SLIT/SPIRACLES: (Fig. 7b). Openings of respiratory tracts. A single pair is found directly behind the *lung slits* in some families whilst other families have a single slit situated between the *spinnerets* and the *epigastric fold*. The spiracles are often difficult to see.

TRICHOBOTHRIUM: (Fig. 8a). Extremely delicate *setal* structure, thinner than a *hair*. Although difficult to see they are important taxonomic features. They are always set vertically in conspicuous but small, round sockets which are often possible to see when the trichobothria themselves are impossible to observe. Varying the angle or illumination may also help to view these characters.

TROCHANTER: (Fig. 7a). The second innermost segment of the *legs* and *pedipalps*.

TUBERCLE. A small rounded projection or protuberance of the *cuticle*.

VENTRAL. The lower or under surface.

Systematic List

The following is a list of the spider families found in Britain. It follows the check list published by Merrett, Locket and Millidge (1985) which is currently accepted by most workers as being definitive. Each scientific name is followed by the English name. Descriptions of each family are given on the pages indicated.

ORDER ARANEAE—Spiders

Orthognatha
    Family ATYPIDAE—Purse Web Spiders 390

Labidognatha
    Family ERESIDAE—Ladybird Spider...........................391
    Family AMAUROBIIDAE—Lace-webbed Spiders 399
    Family DICTYNIDAE—Mesh-webbed Spiders 400
    Family ULOBORIDAE—Cribellate Orb-weavers 400
    Family OONOPIDAE—Six-eyed Spiders 395
    Family DYSDERIDAE—Six-eyed Spiders........................396
    Family SEGESTRIIDAE—Six-eyed Spiders.....................396
    Family SCYTODIDAE—Spitting Spiders 395
    Family PHOLCIDAE—Cellar Spiders or Daddy-Long-Legs Spiders...394
    Family ZODARIIDAE—Ant-eating Spider 393
    Family GNAPHOSIDAE—Ground Spiders .... 407 & 408
    Family CLUBIONIDAE—Foliage Spiders........................408
    Family LIOCRANIDAE—Running Foliage Spiders 408
    Family ZORIDAE—Ghost Spiders 397
    Family ANYPHAENIDAE—Buzzing Spider 406
    Family EUSPARASSIDAE—Green Spider.......................402
    Family THOMISIDAE—Crab Spiders 402
    Family PHILODROMIDAE—Running Crab Spiders...............403
    Family SALTICIDAE—Jumping Spiders 391
    Family OXYOPIDAE—Lynx Spider.............................393
    Family LYCOSIDAE—Wolf Spiders............................398
    Family PISAURIDAE—Nursery-web and Raft Spiders 398
    Family ARGYRONETIDAE—Water Spider.....................406
    Family AGELENIDAE—Funnel-web or Cobweb Spiders.............405
    Family HAHNIIDAE—Lesser Cobweb Spiders.....................405
    Family MIMETIDAE—Pirate Spiders 404
    Family THERIDIIDAE—Comb-footed Spiders 412
    Family NESTICIDAE—Comb-footed Cellar Spider 411
    Family TETRAGNATHIDAE—Long-jawed Orb-weavers...........414
    Family METIDAE—Orb Weavers 414
    Family ARANEIDAE—Orb Weavers 417
    Family THERIDIOSOMATIDAE—Ray Spider 416
    Family LINYPHIIDAE—Money Spiders........................416

NOTES
(key continued overleaf)

388   L. M. JONES-WALTERS

How to Use the Keys

Two keys are included in this guide, each is discussed separately below.

The dichotomous key (p. 390)

The first of the two keys is presented in a conventional *dichotomous* format. Each couplet has two leads, marked *a* and *b*, which offer contrasting descriptions. Choose the lead which corresponds to the specimen being examined and this will direct you to another couplet or provide you with the name of the family to which the specimen belongs.

If in doubt about the presence or absence of a character it should be assumed that it is absent.

Notes are provided for each family. These will review the distinctive morphological features for the group, but also provide ecological notes which may be of use for confirmation purposes. The notes also include cross-references with other texts which will allow you to take identification to specific level if required.

LM&M = Locket and Millidge (1951, 1953), and Locket, Millidge and Merrett (1974). Roberts = Roberts (1985, 1987).

The lateral key (p. 420)

The next key, a *lateral* guide, may prove particularly useful if the specimen is incomplete or damaged, if characters are very difficult to see or if extensive ecological notes are available—on web structure, egg sac carrying, etc.

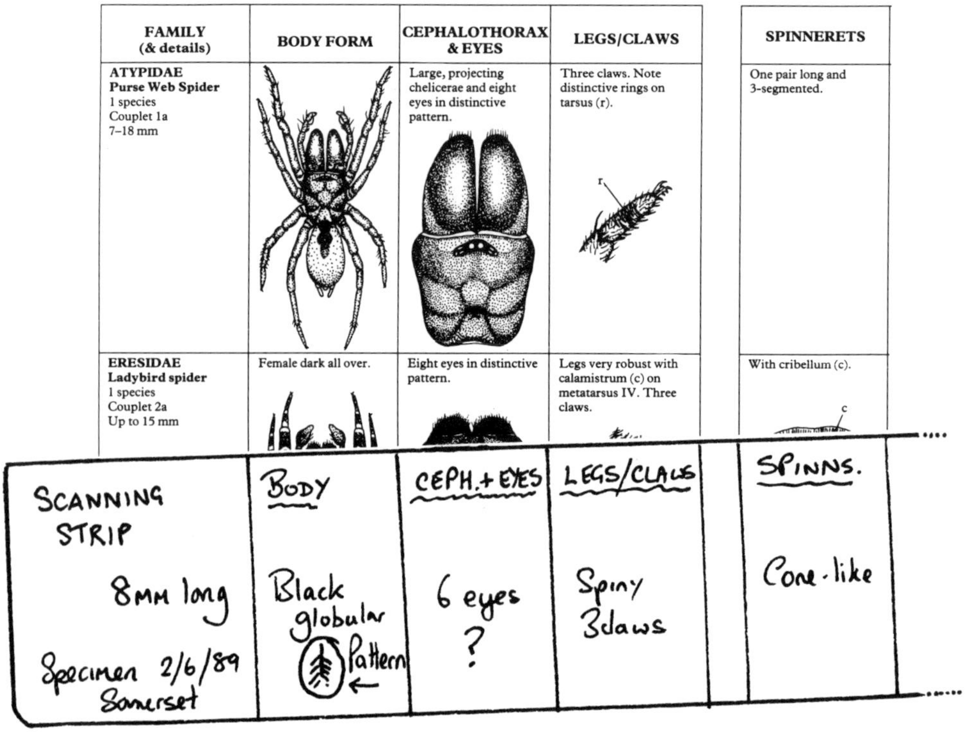

FAMILY (& details)	BODY FORM	CEPHALOTHORAX & EYES	LEGS/CLAWS	SPINNERETS
ATYPIDAE Purse Web Spider 1 species Couplet 1a 7–18 mm		Large, projecting chelicerae and eight eyes in distinctive pattern.	Three claws. Note distinctive rings on tarsus (r).	One pair long and 3-segmented.
ERESIDAE Ladybird spider 1 species Couplet 2a Up to 15 mm	Female dark all over.	Eight eyes in distinctive pattern.	Legs very robust with calamistrum (c) on metatarsus IV. Three claws.	With cribellum (c).
SCANNING STRIP 8mm long Specimen 2/6/89 Somerset	BODY Black globular Pattern	CEPH.+EYES 6 eyes ?	LEGS/CLAWS Spiny 3 claws	SPINNS. Cone-like

Fig. 10.
A scanning strip.

1. The most efficient way to use the table is to manufacture a "scanning strip". This allows all the information for each family to be used simultaneously. The strip should be at least as wide as the table (see Fig. 10), and sufficiently deep to allow you to enter information for your own specimen. In this case the strip will be sub-divided into eight sections, each of which will correspond with the rows containing information for the eight characters—body form, cephalothorax, eyes, etc. A plain formica or plastic strip, used with an indelible ink pen, may prove particularly useful, and will last for several years.
2. The information for the specimen to be identified is then entered in the appropriate sub-section of the strip, and as many sub-sections as possible are filled in. However, not all characters need to be considered and this is where the tabular key will have an advantage over the dichotomous version.
3. The strip is then moved down the table, one family at a time, until a family is reached for which the information recorded on the strip and that given in the table correspond exactly. This name is recorded, but *you then continue* until the bottom of the table has been reached and all families have been "scanned". You may end up with one name, or several alternatives. Either way, the identification should be accepted only after referring to the more detailed descriptions given in the main key, and in other texts. Sometimes, the information entered on the strip may fail to correspond exactly with any family—disagreeing with that entered in the table for one, or more, characters. In this case it will be worthwhile following up the family which appears to correspond *most closely* with your description.

REFERENCES

BRISTOWE, W. S. (1958). *The world of spiders*. New Naturalist Series. Collins, London.

CHALMERS, N. and PARKER, P. (1989). *The OU Project Guide*. 2nd Ed. Occasional Publication of the Field Studies Council No. 9.

COUNTRYSIDE COMMISSION (1985). *Out in the country: where you can go and what you can do*. Countryside Commission, Cheltenham.

FOELIX, R. F. (1982). *The biology of spiders*. Harvard University Press, London.

JOINT COMMITTEE FOR THE CONSERVATION OF BRITISH INSECTS (Undated). *A Code for insect collecting*. Undated pamphlet.

JONES, R. (1983). *Spiders of Britain and Northern Europe*. Country Life Books, London.

LOCKET, G. H. and MILLIDGE, A. F. (1951 & 1953). *British spiders*. Volume I & II. Ray Society, London.

LOCKET, G. H., MILLIDGE, A. F. and MERRETT, P. (1974). *British spiders*. Volume III. Ray Society, London.

MERRETT, P., LOCKET, G. H. and MILLIDGE, A. F. (1985). A check list of British spiders. *Bulletin of the British Arachnological Society* 6(9): 381–403.

MURPHY, F. (1980). *Keeping spiders, insects and other land animals in captivity*. Bartholomew, Edinburgh.

OXFORD, G. S. (1981). An easily constructed holding device for the examination of live spiders. *Bulletin of the British Arachnological Society* 5(6): 278–279.

ROBERTS, M. J. (1985 & 1987). *The spiders of Great Britain and Ireland*. 3 Volumes. Harley Books, Colchester.

SOUTHWOOD, T. R. E. (1978). *Ecological methods: with particular reference to the study of insect populations*. Chapman and Hall, London.

THE DICHOTOMOUS KEY

1a Chelicerae projecting forward horizontally and, when seen from above, these are as long as the cephalothorax (Fig. 11a). Fangs lying along underside of the chelicerae. The posterior spinnerets long, with three segments (Fig. 11b). Two pairs of book lungs visible on the underside of the abdomen (Fig. 11b) . . . . . .Family ATYPIDAE

ATYPIDAE
This family contains a single British Species, Atypus affinis, the purse web spider (Fig. 11c). The adults are large and robust and, in combination with the characters described above, are distinctive. The spider constructs a closed, silken tube most of which is buried in soft earth or detritus (Fig. 11d). Insects crossing the exposed part of the tube are trapped by the spider waiting within. It spears them with its long fangs and pins them against the tube wall which is then slowly cut by the specially-adapted chelicerae. They are then dragged inside and consumed. Purse web spiders may take up to four years to reach maturity. The males become adult in September and October and soon mate with a receptive female. The male may then remain in the female web for several months and there is evidence to suggest that females and even some males may mate and reproduce for more than a single season. Eggs are laid in July and August. Usually found in grasslands, heathlands or mature dune systems. Length: 7–18 mm.
(LM&M Vol I p. 46, Vol III p. 1; Roberts Vol I p. 46, plt. 1)

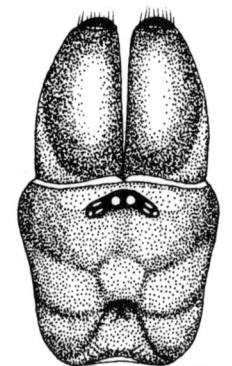

Fig. 11a.

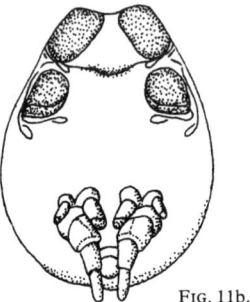

Fig. 11b.

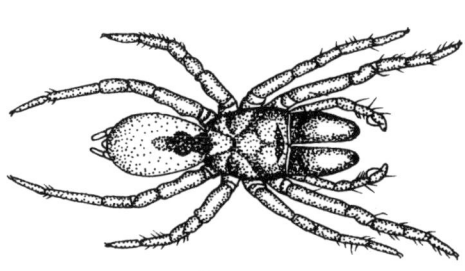

Fig. 11c.

Fig. 11d.

1b Chelicerae projecting downwards, or downwards and forwards, but never horizontally; never as long as the cephalothorax when viewed from above. If chelicerae projecting then fangs are visible from above. Posterior spinnerets never with three segments. Only one pair of book lungs on the underside of the abdomen 2

Keys to the Families of British Spiders 391

2a Body densely clothed with short hairs. Adult extremely large and robust—up to 15 mm. Female completely black, male with a red abdomen with four large black spots (Fig. 12a). Eyes with a characteristic arrangement. Very rare. . . Family ERESIDAE

Fig. 12a.

ERESIDAE
The ladybird spider, Eresus niger, *is the single British representative of the genus. The adults are large and robust and the male is particularly distinctive, his abdominal pattern giving the species its English name. The web consists of a buried silk-lined tube up to 8 cm deep. Above ground one side of the tube lip is extended to form a roof-like structure from which further threads extend to the adjacent vegetation (Fig. 12b). This forms a small sheet in which ground-running invertebrates are ensnared. Mating, which is preceded by a brief courtship, takes place in late April or May, eggs are laid in May, and the life cycle is completed over three to four years. This heathland species is extremely rare and is protected by law. If found it should be left undisturbed. Length: up to 15 mm.*
(LM&M Vol I p. 49; Roberts Vol I p. 46, plt. 2)

Fig. 12b.

2b Not fitting the description above 3

3a The front row of four eyes all facing forwards on a vertical face, much bigger than all the other eyes. The medians are the largest (giving the appearance of car headlamps). The second row of two eyes very small, and often difficult to see. The third row consisting of two medium sized eyes (Fig. 13a) . . . . Family SALTICIDAE

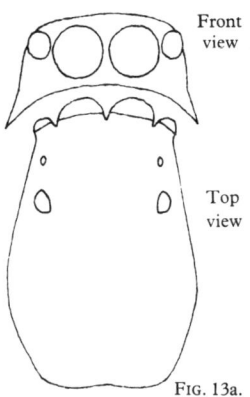

Front view

Top view

Fig. 13a.

SALTICIDAE
The jumping spiders, with 34 species in 15 genera. With their short, robust legs, an almost rectangular cephalothorax with a bank of four huge forward-facing eyes, and a neat abdomen, members of this family are unmistakable (Fig. 13b). The abdomen and cephalothorax are often patterned and two genera, Myrmarachne *and* Synageles *are very impressive ant mimics. There is no web and they are hunters, feeding on other invertebrates. Remarkably, they are able to jump considerable distances on to prey or when escaping danger, which gives them their name. Courtship is complex due to the well developed eyesight, the males flashing semaphore messages to the females with their legs and palps which are often contrastingly marked. Mating takes place in the spring, egg laying in June and July, and the full life cycle is usually completed in a year, although one or two species may take two. They are found in a variety of habitats including grasslands, heathlands, wetlands, scrub and woodland; on the ground, walking over foliage or on bark or walls. Length: 2–10 mm.*
(LM&M Vol I p. 206, Vol III p. 26; Roberts Vol I p. 115, plts. 55–70)

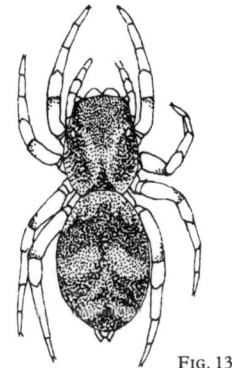

3b Eyes not arranged like this 4

Fig. 13b.

4a Head modified with lobes, turrets, or globular extensions (Fig. 14). Never larger than 3.5 mm . . . .
Some adult males of the Family LINYPHIIDAE
(see page 416 for a full description)

LINYPHIIDAE
A small number of adult males of this family have highly modified head areas, a selection of which are illustrated in Fig. 14.

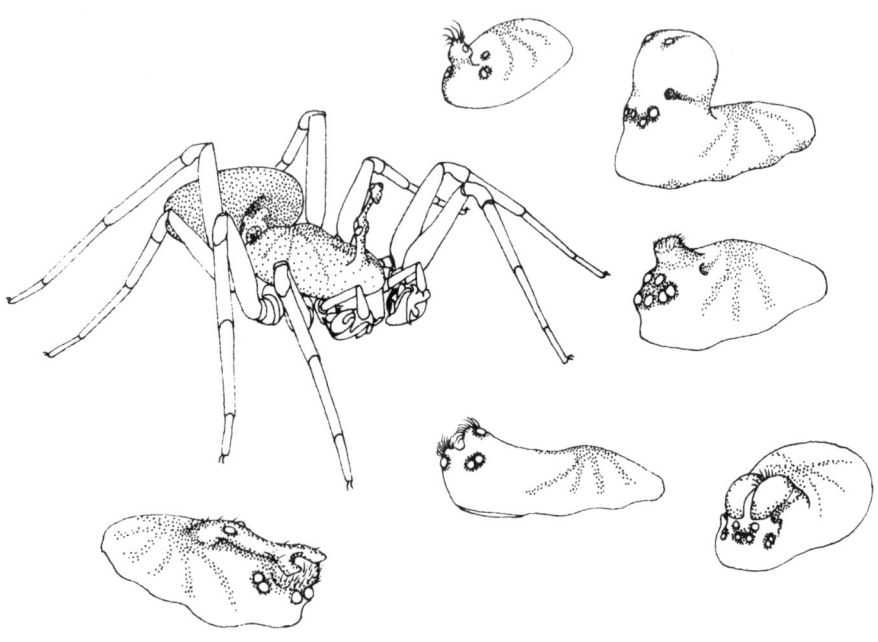

Fig. 14.

Some examples of the head areas of some linyphiid males.

4b Head area, although it may be raised, not modified as extensively as shown in Fig. 14 5
Note: If in doubt, choose the second lead (i.e. go to couplet 5). "Doubtfuls" key out both ways

Keys to the Families of British Spiders

5a Eyes arranged in a hexagonal pattern (Fig. 15a). Legs armed with numerous long, robust spines (more noticeable on living spiders). Abdomen tapering towards the spinnerets (Fig. 15b) . . . . . Family OXYOPIDAE

OXYOPIDAE
Our single species, Oxyopes heterophthalmus, *belongs to this widespread family commonly called the lynx spiders. The hexagonal eye pattern is distinctive and the heavily-spined legs and stance when alive, with the front femora drawn back, are also characteristic. The cephalothorax and abdomen, which tapers towards the spinnerets, are patterned. They catch their, often large, insect prey while running through heather. Mating is in May and early June and eggs are laid in June. The male approaches the female with his palps waving and his abdomen vibrating, raising his front legs in the air before his final advance. The full life cycle takes a year. The species is known from a handful of heathland sites in southern England and it should not be collected unless absolutely necessary. Length: 5–8 mm.*
(LM&M Vol I p. 224; Roberts Vol I p. 130, plt. 71)

5b Eyes and appearance not like this 6

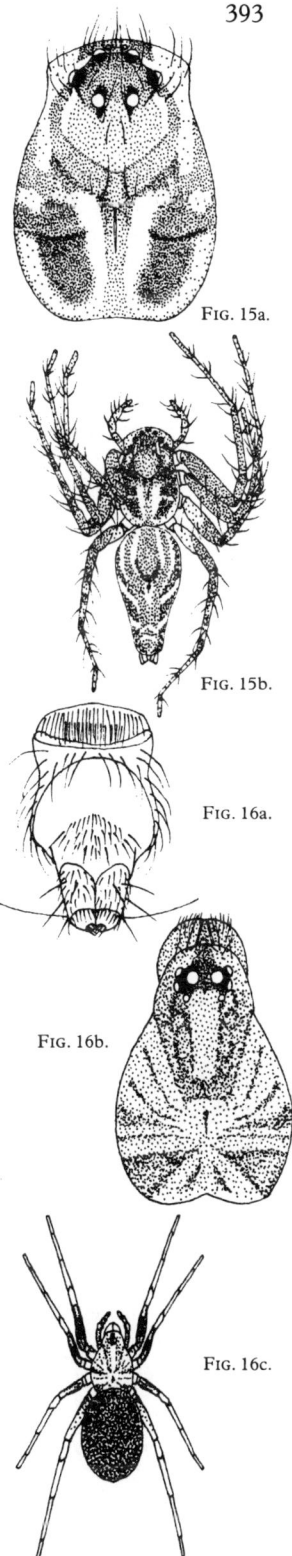

FIG. 15a.

FIG. 15b.

FIG. 16a.

6a Abdomen, when viewed from below, has apparently only a single pair of tube-like spinnerets (Fig. 16a). (When viewed from above the much reduced median and posterior spinners are just visible.) Eye pattern, with large, circular anterior medians, characteristic (Fig. 16b) . . . . . Family ZODARIIDAE

ZODARIIDAE

FIG. 16b.

This family contains a single British representative, Zodarion italicum, *the ant-eating spider (Fig. 16c), which was only discovered in 1985. The tube-like anterior spinnerets and much-reduced median and posterior spinnerets are very distinctive, as are the eyes. The abdomen is also characteristic, being dark brown above and contrastingly pale yellowish below. This is a ground-running species, usually found in the company of the black ant,* Lasius niger, *on which it feeds. The life history is not well known, but is probably completed within a year. It is common in the Grays area of Essex, in chalk pits, but has also been found on the Isle of Grain and may be more widespread. There are reports of a second species having been found on the south coast recently. Length: 2–3 mm.*
(Roberts Vol II p. 172, plt. C)

FIG. 16c.

6b At least two, usually three, pairs of spinnerets visible when the abdomen is viewed from below. Eye pattern not as described above 7

7a Legs very long and spindly (Fig. 17a). Eyes with a characteristic arrangement (Fig. 17b). . Family PHOLCIDAE

PHOLCIDAE
There are 2 British species within this family, Pholcus phalangoides *and* Psilochorus simoni, *which are known as cellar or daddy-long-legs spiders. They both have very distinctive, long, fragile-looking legs with a characteristic eye pattern and almost circular cephalothorax.* Pholcus *has a tubular, grey abdomen,* Psilochorus *a globular, bluish abdomen. There is a tangled, open web (Fig. 17c). A variety of insects and other spiders, often considerably larger than the spider itself, are accepted as prey.* Pholcus *is not averse to leaving its own web and preying on the occupants of the webs of other unrelated species, using a method very similar to that of the Mimetidae (see p. 404). Mating is in the summer and eggs are laid between late May and August. The male enters the female web slowly, gently vibrating his body, finally embracing her and inserting his palps. The life cycle takes one to two years depending on climate, and females of* Pholcus *may survive for more than a year as adults. Both species are usually confined to buildings or outhouses and have a markedly southern distribution. Length: 2–10 mm.*
(LM&M Vol I p. 92, Vol III p. 3; Roberts Vol I p. 64, plts. 18 & 19)

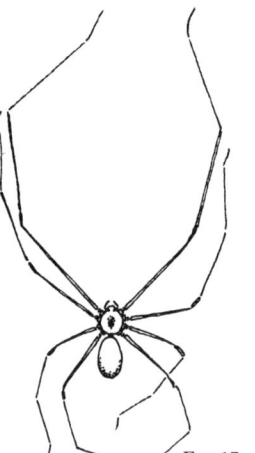

FIG. 17a.

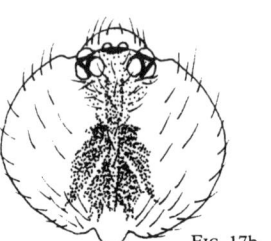

FIG. 17b.

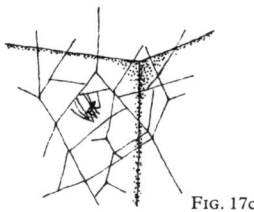

FIG. 17c.

7b If legs long then the eye pattern is not as described above, appearance otherwise not as above 8

8a Six eyes (e.g. Figs. 18b, 19a & 22a) 9

8b Eight eyes (e.g. Figs. 23a, 25a & 26a). 12
Note: Look carefully—some lateral pairs of eyes may touch, to give the appearance of six eyes, not the eight which are actually present.

9a Legs long and slender, body shape distinctive, with a characteristic pattern (Fig. 18a). Eyes arranged as illustrated in Fig. 18b . . . . . Family SCYTODIDAE

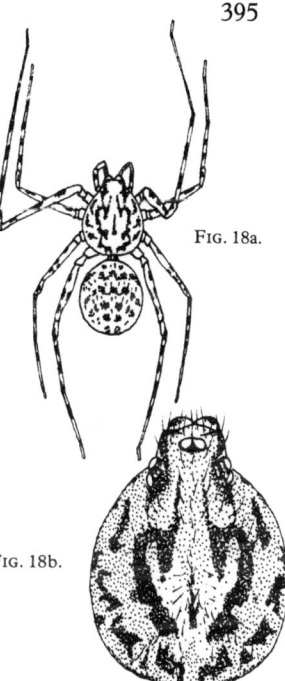

FIG. 18a.

SCYTODIDAE

The single British species, Scytodes thoracica, *is known as the spitting spider. The appearance, with a dark, mottled pattern all over the light brown body, the slender legs, and the domed cephalothorax, is extremely distinctive. There is no web and the species, which wanders slowly over walls and ceilings in houses, has a remarkable method of prey capture. When it encounters an insect a sticky, viscose thread is shot from each chelicera, literally "sticking" the prey, which may be any flying or walking insect of suitable size, to the substrate. Once trapped, the hapless victim is sucked dry. Mating occurs, without courtship, between March and October but egg laying is confined to July and August. The life cycle may take up to three years for completion. Only found in houses in the south of England. Length: 3–6 mm.*
(LM&M Vol I p. 89; Roberts Vol I p. 58, plt. 14)

FIG. 18b.

9b Legs and body shape not as above. Eyes not arranged as above 10

10a Adult body length not more than 3 mm. Median eyes slightly larger than laterals and arranged as illustrated in Fig. 19a. Two claws (without claw tufts), held on a flexible, chitinous extension called an onychium (Fig. 19b). Labium as wide as long . . . . Family OONOPIDAE

FIG. 19a.

OONOPIDAE

One of the families called six-eyed spiders; we have 2 native species, both in the genus Oonops. *They are extremely small and distinctive, being coloured pink all over. They weave no web and rely on stealth to catch their prey. Hunting at night, they catch a variety of small flies and other insects. O. pulcher is occasionally found scavenging in the webs of amaurobiids and agelenids, feeding on the remnants of their meals. The mating ceremony is brief; the male stroking the female's legs with his own, the female snapping her chelicerae. The life cycle is probably a year but eggs are laid throughout the summer months and adults may be found in every season. O. domesticus is usually encountered in houses while* O. pulcher *is found outside, under stones, bark, dry grass tussocks and sometimes birds' nests. Length: Up to 2 mm. (Note: There are three other alien oonopids found in Britain, all of them from the tropics, which survive in hothouses at Kew and elsewhere. They are also tiny, but are generally brown and have hard sclerotised plates (scuta) covering part of the upper and lower surfaces of the abdomen. One species,* Diblemma donisthorpei *has only two eyes).*
(LM&M Vol I p. 73; Roberts Vol I p. 58, plt. 13)

FIG. 19b.

FIG. 20.

10b Adults considerably larger than 3 mm. Eyes not arranged as above. Tarsi either with three claws or with two claws and tufts (Fig. 20). Labium much longer than wide . . . 11

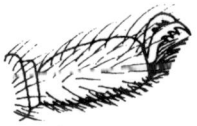

11a Eyes grouped in a cluster forming a circle (Fig. 21a,b). Abdomen without a defined pattern (Fig. 21c). . . . .
Family DYSDERIDAE

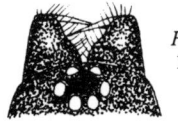

Harpactea
FIG. 21a.

DYSDERIDAE

The second family of six-eyed spiders, containing 3 species and 2 genera, Dysdera *and* Harpactea. Dysdera *can be very large and has a very characteristic deep red cephalothorax and legs with a pinkish grey abdomen (Fig. 21c). The huge, jutting jaws of this genus, which bear a pair of savage-looking fangs, are also distinctive (Fig. 21b). They are ground hunting species and, during the day, are usually found in silk cells under stones or amongst thick clumps of vegetation, emerging at night to search for prey. Unusually among predatory arthropods, they feed exclusively on woodlice, for which their jaws are specially adapted. Mating, which involves little ceremony, is in the spring and the female lays eggs in June and July. It is likely that the full life cycle takes at least two years for completion. Found in grasslands, heathlands and gardens. In* Harpactea hombergi, *the cephalothorax and legs are dark brown and the distinctly tubular abdomen varies from pale brown to dark grey. There is no web and, like* Dysdera, *this species feeds at night, retreating to a silken cell during the day. The prey is often other spiders, but other ground-running invertebrates are taken. Following mating, eggs are laid from May to July, and the life cycle may be completed in a year. Found under bark and stones, tussocks, occasionally birds' nests, or in houses. Length: 7–15 mm.*
(LM&M Vol I p. 81; Roberts Vol I p. 60, plts. 15 & 16)

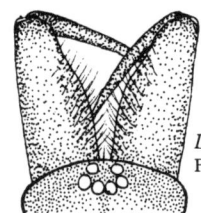

Dysdera
FIG. 21b.

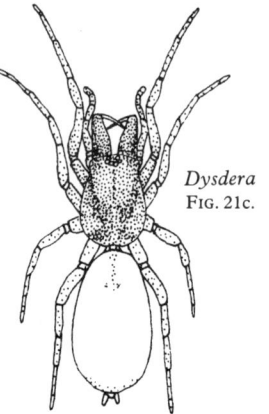

Dysdera
FIG. 21c.

11b Eyes not forming a circle (Fig. 22a). Abdomen usually with a characteristic pattern (Fig. 22b) . . . .
Family SEGESTRIIDAE

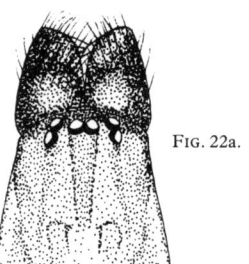

FIG. 22a.

SEGESTRIIDAE

This family, also known as six-eyed spiders, contains 3 British species all of the genus Segestria. *The carapace is dark brown or black and the tubular abdomen varies from grey to black. It often carries a distinctive pattern. The first three pairs of legs are all directed forwards. Segestriids live in silken tubes constructed in holes in walls or under bark and stones. Each tube has up to twelve distinctive threads radiating from the lip which act as trip wires to alert the spider of passing insects, which are attacked and dragged into the retreat (Fig. 22c). Mating is between spring and late summer when the males enter the female webs, and the full life cycle may take two years or more. Two of the species are found only in the south of Britain, but the third is found up to the north of Scotland in walls and tree bark. Length: 6–22 mm.*
(LM&M Vol I p. 81, as g. Segestria; *Roberts Vol I p. 62, plt. 17)*

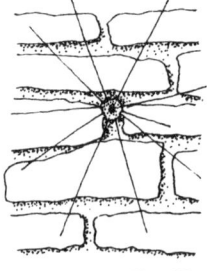

FIG. 22c.

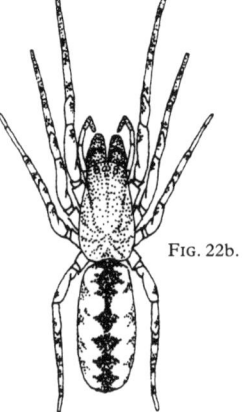

FIG. 22b.

Keys to the Families of British Spiders

12a Eyes WHEN VIEWED FROM ABOVE arranged in three rows. (e.g. Fig. 23a) 13

12b Eyes WHEN VIEWED FROM ABOVE arranged in two rows. (e.g. Fig. 30a) 15

13a All eyes similar in size and when viewed from above, the front row on a sloping face (Fig. 23a). Cephalothorax with a pointed appearance. Background colouration of the body pale, with darker markings (Fig. 23b). Two tarsal claws (Fig. 23c). . . . . Family ZORIDAE

ZORIDAE
This family, the ghost spiders, includes 4 species in the genus Zora. The eye pattern and narrow head area, which gives the cephalothorax a pointed appearance, are characteristic. The body has a pale yellow-brown background colour with darker brown markings and is distinctive. They are hunting species, often found during the day, feeding on a variety of other invertebrates. Mating involves little or no courtship, and takes place in May and June with eggs laid in June and August. The life cycle is probably completed in a year. Found in ground vegetation in grasslands, heathlands and woods, often in damp situations and moss. Length: 2.5–6.5 mm.
(LM&M Vol I p. 156, as g. Zora, Vol III p. 17; Roberts Vol I p. 94, plt. 37)

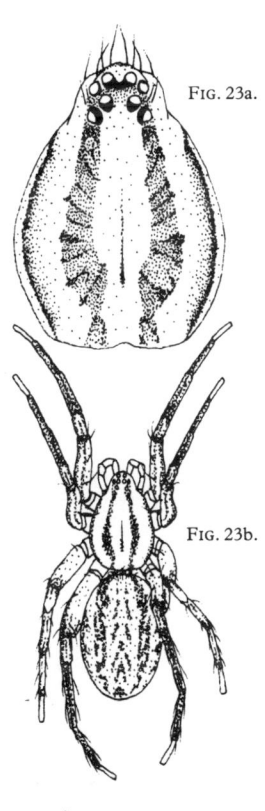

Fig. 23a.

Fig. 23b.

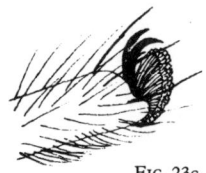

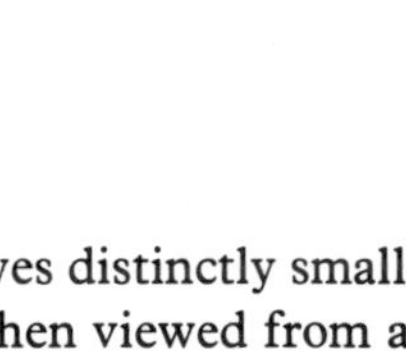

Fig. 23c.

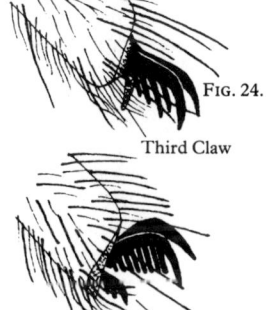

Fig. 24.

Third Claw

Third Claw

13b The front row of four eyes distinctly smaller than the rest, and on a vertical face when viewed from above (e.g. Figs. 25a & 26a). Cephalothorax with a blunt appearance (e.g. Figs. 25b & 26b). Background colouration generally dark. Three tarsal claws (which may be difficult to see unless magnification of 50–100 × is used) (Fig. 24) 14

14a Eyes WHEN VIEWED FROM IN FRONT comprising three rows, with a front row of four small eyes, a larger pair immediately behind and a similar pair, but more widely spaced, lying further back (Fig. 25a) . . . .
Family LYCOSIDAE

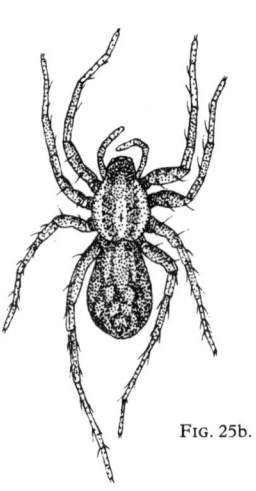

Front view
FIG. 25a.

LYCOSIDAE

The 36 species in 9 genera forming this family are all called wolf spiders. They have a very characteristic arrangement of their large, glassy eyes which are aligned in three rows. Usually dark, or cryptically coloured, the abdomen and cephalothorax often bear a symmetrical pattern (Fig. 25b). Most of the species are ground-running hunters, often found in huge abundance in the summer, feeding on insects and other spiders. Some species are nocturnal, others hunt by day; those of the genus Arctosa *live in silken tubes from which they pounce on passing insects. Mating, which involves much leg waving and palp vibrating on the part of the male, is in the spring and early summer and the female carries the egg sac attached to her spinnerets, which is a distinctive character of the family in the field. Most species complete their life cycle in a year, but large ones may take up to two years. Most species are ground running in marshlands, bogs, grasslands, heathlands and woodlands. One or two may be found on the foliage of trees and bushes. Length: 3.5–18 mm.*
(LM&M Vol I p. 247, Vol II p. 406, Vol III p. 31; Roberts Vol I p. 132, plts. 72–85)

FIG. 25b.

14b Eyes WHEN VIEWED FROM IN FRONT comprising two rows of four eyes each (Fig. 26a) . . . .
Family PISAURIDAE

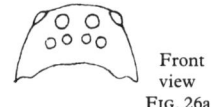

Front view
FIG. 26a.

PISAURIDAE

This family contains 3 species, Pisaura mirabilis, *the nursery web spider, and 2 species of* Dolomedes, *the raft spiders. They are all large and distinctive as adults with a characteristic eye pattern.* P. mirabilis *always has a distinctive light stripe down the centre of the cephalothorax, and a variable pattern on the long, tapered abdomen, which is usually lighter along the flanks (Fig. 26b).* Dolomedes *is dark with two cream lines running along the sides of the cephalothorax which continue along the flanks of the abdomen (Fig. 26c). There is little other patterning on* Dolomedes *which is very robust with thick, strong legs.* Pisaura *is an active ground-running predator, feeding on a variety of, often large, invertebrate species.* Dolomedes *lies in wait on the edge of pools and ditches with a leg touching the water surface, sensitive to any ripples or disturbance in the water. Attracted by movement they rush across the water, seize their prey and drag it to the bank. Larger prey, including minnows, may also be grasped by the spider which, anchored to the bank or a suitable piece of vegetation by its hind legs, hurls itself, prey and all, backwards and out of the water. The prey is rapidly subdued by an extremely effective poison.* Pisaura *mates in May and in a complex courtship ritual the male presents the female with a silk-wrapped insect by way of distraction prior to copulation (Fig. 2). In July, the eggs are laid and carried by the female in her chelicerae until, just before*

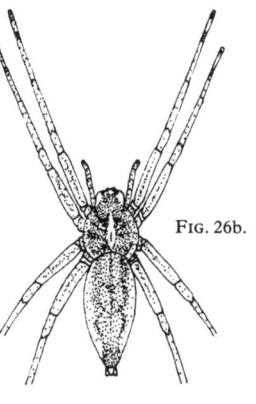

FIG. 26b.

hatching, they are placed in a domed nursery web which she guards prior to the dispersal of the spiderlings. The full life cycle is probably completed in a year. Dolomedes mates in May and June, and constructs a similar nursery web, which may be distinguished from that of Pisaura by its bluish tinge. The eggs which are laid soon after are carried Pisaura-like in the chelicerae. The life cycle is completed in a year but females in particular may live up to eighteen months. Pisaura is found in abundance in grasslands, heathlands and woodland rides, adult Dolomedes around standing water particularly on heathland, the juveniles more often in low bushes and in trees in the surrounding habitat. Dolomedes plantarius, the fen raft spider, is only known from a small area in East Anglia, is protected by law, and if found should not be collected. Length: 9–22 mm.
(LM&M Vol I p. 292, Vol III p. 39; Roberts Vol I p. 154, plts. 86 & 87)

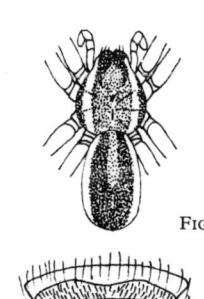

Fig. 26b.

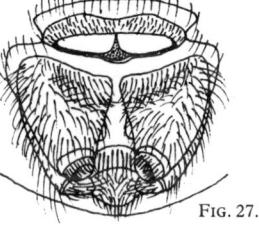

Fig. 27.

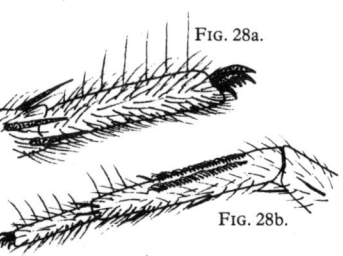

Fig. 28a.

Fig. 28b.

15a With a *cribellum* in front of the spinnerets (Fig. 27), and a *calamistrum* on metatarsus IV (e.g. Fig. 28b). The *cribellum* is always present, the *calamistrum* varies from just a few bristles (which may be extremely difficult to see or absent in males) to a row extending the entire length of the metatarsus 16

15b Without a *cribellum* or a *calamistrum* 18

16a All eyes light in colour. Tarsi with a dorsal row of trichobothria (Fig. 28a). Females with a calamistrum consisting of two rows of bristles (Fig. 28b) . . . .
Family AMAUROBIIDAE

AMAUROBIIDAE
The lace webbed spiders. Containing a single British genus, Amaurobius, *with 3 species. They are relatively large as adults and are all reasonably common. They are quite dark in appearance, with a characteristic, lighter, abdominal pattern (Fig. 28c). The web is a tangled mesh of threads surrounding a circular retreat leading into a crevice, usually located on a vertical surface (Fig. 28d). When freshly spun, the threads have a lace-like appearance and a bluish tinge caused by the action of the cribellum and calamistrum, which combine to produce composite and adhesive silken strands. The prey are usually surface-running invertebrates which stumble across the web. There is a two year life cycle and mating takes place between late August and September, the male approaching the female's web vigorously vibrating his palps and legs; eggs are laid in June and July of the following year. The webs are most commonly found in walls and the bark of trees. Length: 4–15 mm.*
(LM&M Vol I p. 53, as g. Ciniflo, *Vol III p. 1; Roberts Vol I p. 48, plt. 3)*

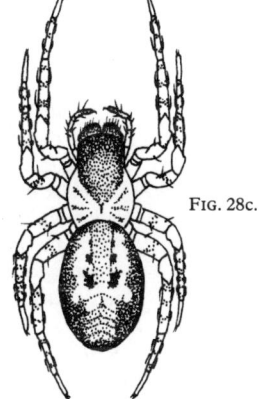

Fig. 28c.

16b All eyes dark, or a combination of light or dark eyes. Tarsi either without trichobothria or with one at most. Females with a calamistrum consisting of a single row of bristles.
. . . . 17

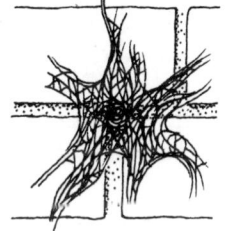

Fig. 28d.

17a All eyes dark with both rows curving backwards (recurved), the posterior row most markedly (Figs. 29a,b). Metatarsus IV compressed and curved under the line of the calamistrum (Figs. 29c,d) . . Family ULOBORIDAE

> **ULOBORIDAE**
> *The cribellate orb weavers. The 2 species in this family,* Uloborus walkenaerius (Fig. 29f) *and* Hyptiotes paradoxus (Fig. 29e) *differ somewhat in appearance. Both are rather uncommon and will not easily be confused with any other species.* U. walkenaerius *weaves a complete orb web which is strung horizontally across low vegetation (Fig. 29h).* H. paradoxus *makes only a triangular section of an orb web (and is sometimes called the triangle spider) (Fig. 29g). Both species capture flying insects.* H. paradoxus *has a remarkable technique of holding the triangle of the web taught until an insect flies into it, then literally "springing" the trap by releasing the web in stages around the struggling prey. There is a one- or two-year life cycle with mating in the spring (* H. paradoxus*) and summer (* U. walkenaerius*). Eggs are laid soon after mating which is a brief ceremony in both species.* U. walkenaerius *is a heathland species,* H. paradoxus *is found on the upper branches of evergreen trees and bushes, usually yew or box. Length: 3–6 mm.*
> *(LM&M Vol I p. 70; Roberts Vol I p. 56, plts. 11 & 12)*

17b The anterior median eyes alone dark, and at least the anterior row almost straight Fig. 30a. Metatarsus IV not compressed and curved under the line of the calamistrum (Fig. 30b) . . . . Family DICTYNIDAE

> **DICTYNIDAE**
> *The mesh webbed spiders, with 15 species in 5 genera (Fig. 30c). Generally very small spiders, often dark, weaving small webs of flocculent or fluffy silk radiating downwards from the top of grasses or tall herbs, or across single leaves of trees and shrubs (Fig. 30d). They feed on any flying or crawling insects that find their way into the webs. The life cycle is usually a year in length and mating in most species takes place in the summer. The male and female spend up to a month sharing the web before eggs are laid. Found in a variety of habitats, including grasslands, heathlands, hedgerows, scrub and woodland edges. Length: Up to 4 mm.*
> *(LM&M Vol I p. 51, Vol III p. 2; Roberts Vol I p. 48, plts. 8–10)*

18a Eyes distinctly black and beady, usually both rows procurved (back row in Eusparassidae very slightly recurved) (e.g. Fig. 31c). Legs I and II turned through 90 degrees so that the surface which is normally at the bottom, faces forwards; often more robust than legs III and IV (e.g. Fig. 31b) 19

18b Eyes clear and glassy, if procurved then not black and beady. Legs not as described above 21

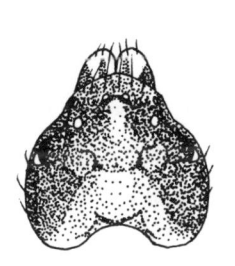

Fig. 29a. 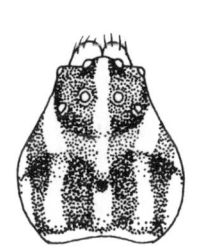   Fig. 29b.    Fig. 29c.    Fig. 29d.

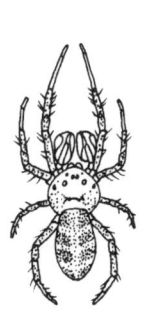

Fig. 29e. 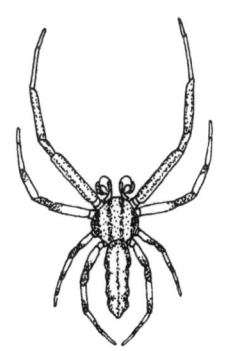   Fig. 29f. 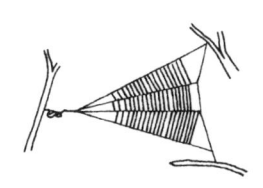   Fig. 29g.    Fig. 29h.

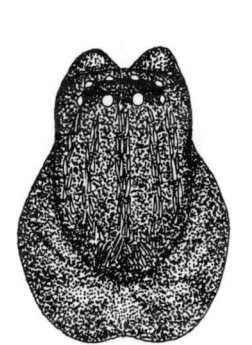

Fig. 30a. 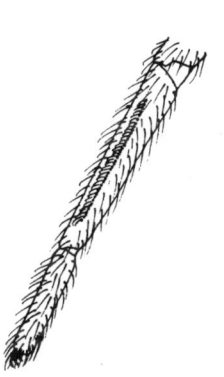   Fig. 30b. 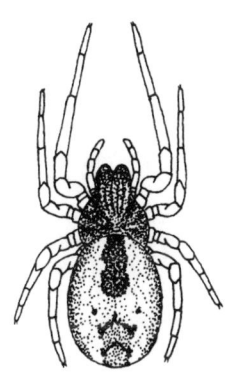   Fig. 30c.    Fig. 30d.

19a Claw tufts absent, or tips of tarsi with simple hairs. Tarsi I and II without scopulae (Fig. 31a). Distinctly crab-like in appearance (Fig. 31b). . . . Family THOMISIDAE

THOMISIDAE
This family contains 25 British species in 6 genera. They are called crab spiders. Their eyes, which are in two procurved rows, are distinctively black and beady (Fig. 31c), the cephalothorax is almost circular when seen from above and the abdomen is dumpy and squat. The first two pairs of legs are longer and more robust than the second pair and the general appearance is crab-like. They are very often brightly coloured or patterned. They have no web and lie in wait on the ground, on vegetation or in the cups of flowers whose colours they may mimic for camouflage. They leap on their invertebrate prey, restraining it with the large front legs which are often armed with spines. Mating and egg laying take place mainly in the spring and early summer and the life cycle is usually completed in a year. In most species there is no courtship and the male mates and leaves. However, in the genus Xysticus *the male ties the female down with silk prior to mating. They are found in a variety of habitats including grasslands, heathlands, wetlands, scrub and woodland; on the ground and walking over foliage. Length: 2–10 mm.*
(LM&M Vol I p. 169, Vol II p. 406, Vol III p. 21; Roberts Vol I p. 97, plts. 40–54)

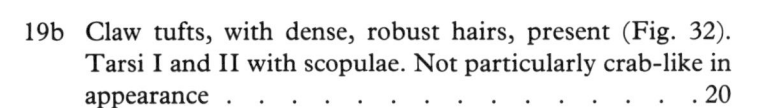

FIG. 31a.

FIG. 31b.

FIG. 31c.

19b Claw tufts, with dense, robust hairs, present (Fig. 32). Tarsi I and II with scopulae. Not particularly crab-like in appearance 20

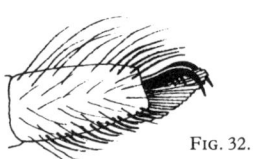

FIG. 32.

20a Eyes ringed with conspicuous white hairs (Fig. 33a). Adult females bright green all over, adult males with green legs and cephalothorax and a distinctive yellow abdomen with a bright red central stripe and red sides. Juveniles brownish all over, with fainter white rings around eyes, abdomen with two pale yellow stripes. Rear margin of the cheliceral fang furrow armed with teeth. Face of each chelicera with a "fan" of stiff hairs in front of the fangs (Fig. 33b) . . .
Family EUSPARASSIDAE

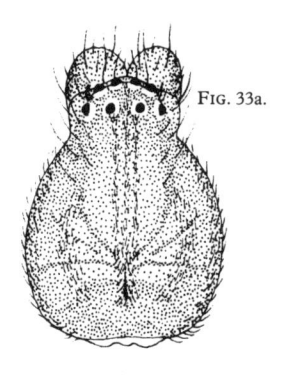

EUSPARASSIDAE
The one British species, Micrommata virescens, *is most appropriately called the green spider. As an adult it is very distinctive, the female is coloured bright green all over, the male has green legs and cephalothorax and a yellow abdomen with a red central stripe and red or green flanks (Fig. 33c). The juveniles are hay coloured and all ages have rings of conspicuous white hairs around their eyes. Preserved specimens tend to fade to a uniform yellow and lose their eye rings. They have no web and hang downwards in grassy vegetation ready to spring on passing insects. The joint between the tarsus and metatarsus on the first two pairs of legs has a soft membrane which permits increased extension and flexibility of the legs allowing*

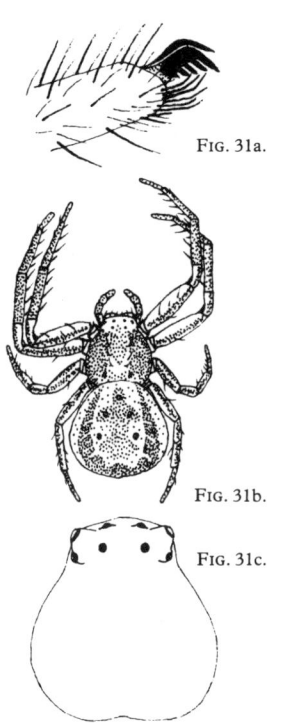

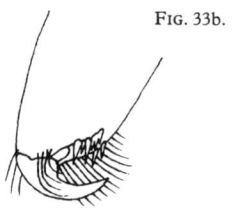

FIG. 33a.

FIG. 33b.

particularly large and vigorous prey, like grasshoppers, to be held and subdued during their struggles. Mating is in June and July and eggs are laid soon after. The male simply leaps on the female, grabbing her in his chelicerae prior to inserting his palps. The life cycle is completed in a year. They are found in low vegetation, more often in the south, often in damp sheltered spots along rides and in clearings in woodland. Purple moor grass, Molinia caerulea, is a favoured habitat, particularly on damp partly wooded heathlands. Length: 7–13 mm.
(LM&M Vol I p. 166, as f. Sparassidae, Vol III p. 20; Roberts Vol I p. 96, plt. 39)

FIG. 33c.

Note: After preservation in alcohol this species loses its green colour and both males and females turn uniformly yellow and lose the white eye rings; the teeth at the rear of the cheliceral fang furrow should then be referred to. A single thomisid species has green legs and cephalothorax, with white rings around the eyes, and could be confused with *Micrommata*; however, it has a dull brown abdomen and lacks claw tufts and scopulae and teeth on the rear margin of the fang furrow.

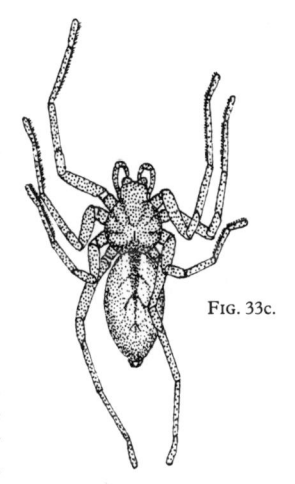

20b Eyes not as above. Adults and juveniles brown, black or mottled, never green. Without teeth on the rear margin of the cheliceral fang furrow. Face of each chelicera without a "fan" of stiff hairs in front of the fangs . . . .
Family PHILODROMIDAE

PHILODROMIDAE
Known as the running crab spiders and closely allied to the Thomisidae, this family contains 15 species in 3 genera. The eyes are black and beady, in two procurved rows, and the cephalothorax is almost circular when seen from above (Fig. 34a). The abdomen, although large and rounded in egg-bound females, is not noticeably dumpy and squat—and is long and tubular in certain genera (Fig. 34b). They are generally brownish and may have a cryptic patterning. The legs are nearly uniform in length. They are active hunters of other invertebrates and are capable of extremely rapid movement when necessary. Mating, which is without courtship, is in spring and early summer and is quickly followed by egg laying. The full life cycle is usually completed in a year. They are found in a variety of habitats including grasslands, heathlands, dune systems, scrub and woodland; on the ground and walking over foliage. Occasionally, they are found in buildings. Length: 4–7 mm.
(LM&M Vol I p. 194, as g. Philodromus, Thanatus, Tibellus, Vol III p. 23, as g. Philodromus; Roberts Vol I p. 108, as g. Philodromus, Thanatus, Tibellus, *plts. 51–54)*

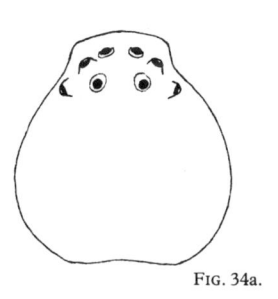

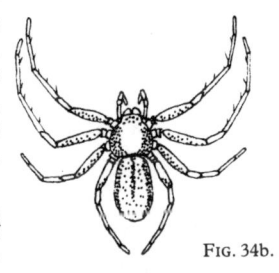

FIG. 34a.

FIG. 34b.

21a Tibia and metatarsus I and II with a forward-facing row of long spines, between which are single rows of shorter spines, curved near their tips and increasing in length towards the end of the row (Fig. 35b) (look carefully). Up to 4 mm . . . . Family MIMETIDAE

MIMETIDAE
Called the pirate spiders, with 4 species all in the genus Ero. *The first two pairs of legs of these small, globular spiders are characteristically spiny and distinctive* (Fig. 35a). *In addition, the upper surface of the abdomen often bears one, two or three pairs of tubercles. They are wanderers, feeding on other spiders of the families Theridiidae and Linyphiidae, and spin only a temporary web of a few threads. They enter the webs of their unsuspecting hosts by stealth, and once in a suitable position mimic the struggles of a prey item by jerking at the threads around them. The occupant is then lured to its doom. Mating is brief and takes place in the spring, eggs are laid in the summer and the life cycle is probably completed in a year. Usually found in grasslands and heathlands. Length: 2.4–4.0 mm.*
(LM&M Vol II p. 32, Vol III p. 46; Roberts Vol I p. 170, plts. 102 & 105)

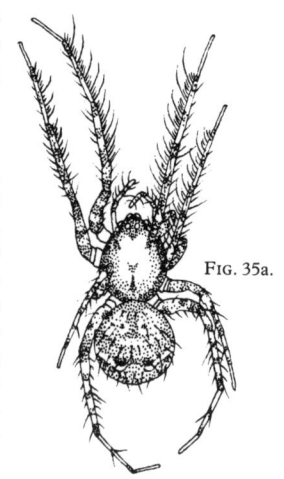

FIG. 35a.

FIG. 35b.

21b Tibia and metatarsus I and II either without spines or with a spinal arrangement other than that described above. Various sizes–but sometimes larger than 4 mm . . . 22

22a One pair of spinnerets clearly two-segmented (e.g. Fig. 36a & 37b). Tarsi with a single row of trichobothria (e.g. Fig. 37d) 23

Note: In some species the two-segmented spinnerets are not always obvious, particularly to the inexperienced eye (Fig. 37a). The trichobothria should always be referred to at this point (Fig. 37d).

22b Spinnerets not obviously two segmented. Tarsal trichobothria, if present, not as above 24

23a The six spinnerets in a more-or-less transverse row (Fig. 36a) . . . . Family HAHNIIDAE

HAHNIIDAE
Closely related to the agelenids (see below), this family contains 7 British species in 2 genera, Hahnia and Anistea: the lesser cobweb spiders. The transverse row of spinnerets, the outermost of which is two-segmented, are very characteristic. The abdomen may be marked with a simple pattern which can be obscure in darker species (Fig. 36b). They spin a small sheet web without an obvious funnel retreat and prey on small insects which enter the web. Mating and egg laying is probably in the spring and summer but adults of some species may be found all year round. The life cycle is likely to be completed within a year. They are usually found close to the ground in webs strung between fissures in the soil, stones, moss, leaf litter and other vegetation (Fig. 32c). Often associated with damp habitats. Length: 1.3–3 mm.
(LM&M Vol II p. 26, as g. Anistea, Hahnia, Vol III p. 45 as g. Hahnia; Roberts Vol I p. 168, plts. 97–101)

FIG. 36a.

FIG. 36b.

FIG. 36c.

23b Spinnerets not arranged in a row, but appear as illustrated in Fig. 37a,b. . . . . . . Family AGELENIDAE

AGELENIDAE
This family, the funnel-web or cobweb spiders, contain 17 species in 8 genera. The two-segmented posterior spinnerets are characteristic; as is the single row of trichobothria on the tarsus of each leg which increase in length towards the end of the leg (Fig. 37d). The abdomen is usually patterned and the cephalothorax may also be distinctively marked (Fig. 37e). These spiders build a sheet web or cobweb of varying size which, on one edge, is rolled up to form a funnel-shaped retreat (Fig. 37c). They prey on crawling and flying insects which find their way onto the top of the web or walk across the outer threads, dragging them into the funnel retreat to be consumed. Mating takes place in spring and summer depending on the species and egg sacs are laid soon after. There is no elaborate courtship, although the males may vibrate the web prior to approaching the female. Some Tegenaria species may take up to three years to reach maturity, and females may live to mate and lay eggs in consecutive years. The webs may be found in tree bark, roots, mossy banks, walls and in buildings. Adults of many of the large, long-legged Tegenaria species are frequently found running in houses in the summer and autumn. Length: 2–17 mm.
(LM&M Vol II p. 1, Vol III p. 41; Roberts Vol I p. 156, plts. 89–96)

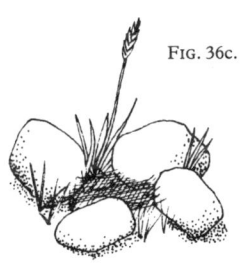

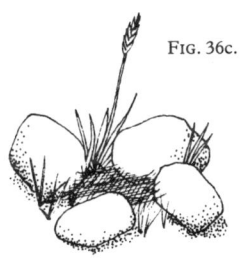

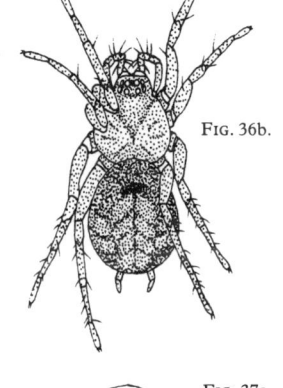

FIG. 37a.

FIG. 37b.

FIG. 37c.

FIG. 37d.

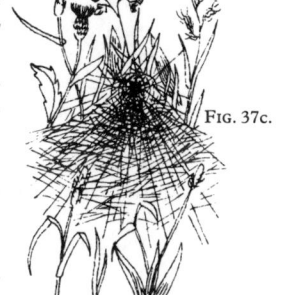

FIG. 37e.

24a Abdomen with the characteristic pattern illustrated in Fig. 38a. Tracheal slit halfway between the spinnerets and the epigastric furrow (Fig. 38b). Family ANYPHAENIDAE

ANYPHAENIDAE

The single British species, Anyphaena accentuata *is called the buzzing spider. The abdominal pattern is characteristic, as is the tracheal slit which is clearly visible between the spinnerets and the epigastric furrow. The species does not spin a web and is usually found in trees and bushes where it runs rapidly over the foliage catching insects. The mating, which takes place in late May, is remarkable, the male emitting a clearly audible buzz as he vibrates his abdomen on the surface of leaves when approaching the female. Eggs are laid in mid-June and the life cycle takes a year or two years for late-hatching broods. Found in individual trees, scrub and woodland. Length: 4–7.5 mm. (LM&M Vol I p. 163; Roberts Vol I p. 94, plt. 38)*

24b Abdomen without the pattern illustrated in Fig. 38a, although it may be otherwise marked. Tracheal slit or spiracle either closer to the spinnerets or closer to the epigastric fold, not situated halfway between. 25

FIG. 38a.

FIG. 38b.

25a Tracheal spiracles, which appear as a single slit, visible behind the epigastric fold (Fig. 39a). Legs III and IV covered with dense, almost tufted hairs (Fig. 39b) . . . Family ARGYRONETIDAE

ARGYRONETIDAE

The single British species, Argyroneta aquatica, *is called the water spider and lives a completely aquatic, submerged existence. The tracheal spiracles, which are easily visible behind the epigastric fold, are characteristic, together with legs III and IV which are clothed with dense, almost tufted hairs and the abdomen which has a thick "pelt" of shorter hair. There are rows of short dark hairs on the cephalothorax. The habitat is also diagnostic. There is a silken bell constructed under water from which threads are trailed. However, this is not used as a snare. This species is a hunter, feeding on a variety of aquatic invertebrates captured as it forages among weeds and other vegetation. Adults may be found throughout the year, but mating occurs in the spring, in the bell without any elaborate courtship, and egg sacs are laid in the summer and early autumn. The completion of the life cycle may take up to two years and adults may live for longer. They are found in fresh water without strong currents all over the British Isles but, due to changing agricultural practices, are becoming increasingly rare. Length: 8–15 mm.*
(LM&M Vol II p. 5, as g. Argyroneta, *Vol III p. 41, as g.* Argyroneta; *Roberts Vol I p. 154, plt. 83)*

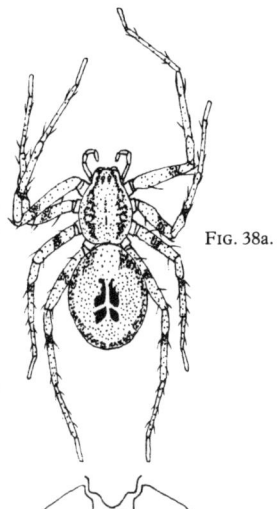

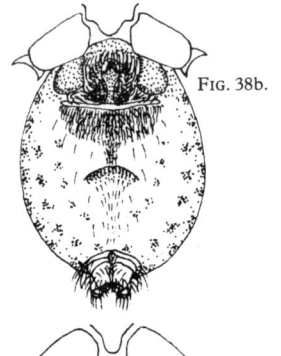

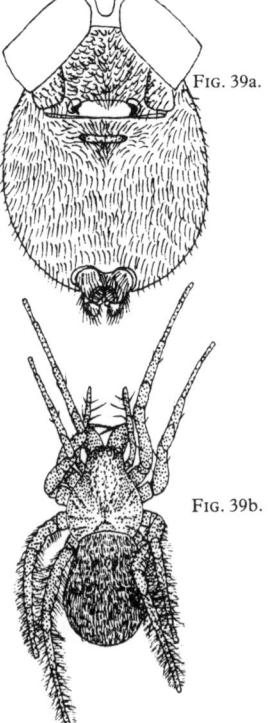

FIG. 39a.

FIG. 39b.

25b Tracheal spiracles, when visible, not situated behind the epigastric fold but close to the spinnerets. Legs III and IV, although they may be hairy, not covered with dense almost tufted hairs 26

26a   Tarsi with two claws, with (e.g. Fig. 41c) or without (e.g. Fig. 44a) claw tufts (see the note below if experiencing difficulty with this character). If tufts are present assume that the tarsus has two claws. Sleek (often with the appearance of having a well-groomed coat, as a result of the short, dense abdominal hairs in many species). The spinnerets nearly always visible at the end of the abdomen when the (usually elongate) body is viewed from above (e.g. Fig. 41e). . . 27

26b   Tarsi with three claws, never with tufts (care should be taken at this point as the third, unpaired, median claw is quite small and often difficult to see) (e.g. Fig. 45b). Abdomen, if hairy, coarse looking, not sleek. The spinnerets not usually visible from above, hidden by the curve of the more rounded abdomen (e.g. Figs. 45c & 46c) 30

Note: In order to see the claw characters in this couplet magnification of between 50 × and 100 × may be required, depending on light and the optical quality of the microscope. If there are problems, a number of other characters are given in order to make the separation easier.

27a   Upper (anterior) spinnerets, when viewing the underside of the abdomen, cylindrical, separated at their bases by a distance about equal to their diameter, and more heavily sclerotised than the lower (posterior) ones (Fig. 40a). Eyes distinctly different in shape, the anterior medians often oblique, oval or triangular . . Family GNAPHOSIDAE

FIG. 40a.

GNAPHOSIDAE
The ground spiders, with 33 British species in a number of genera. Their sleek, elongate but robust bodies which lack any well-defined pattern (Fig. 40b), *the well-developed legs and the distinctly tubular spinnerets (which, in life, take on the appearance of a mobile hand and fingers, when paying-out thread), combine to distinguish the family. The family as a whole are mainly ground-running, nocturnal hunters which feed largely on other spiders. Adults may be found in most seasons and maturity may be reached in one or two years depending on the species. Mating involves fierce sparring that may result in the death of the male or the female, and egg laying takes place in the spring and summer. Found under stones and bark or in grasslands, heathlands and on woodland floors, occasionally in houses. Length: 4–18 mm.*
(LM&M Vol I p. 95, Vol II p. 406, Vol III p. 5; Roberts Vol I p. 64, plts. 20–27)

FIG. 40b.

27b   Upper (anterior) spinnerets, when viewing the underside of the abdomen, cone shaped, hardly separated at their bases, and not more heavily sclerotised than the lower (posterior) ones (e.g. Fig. 41d). All eyes of a similar size, shape and colour 28

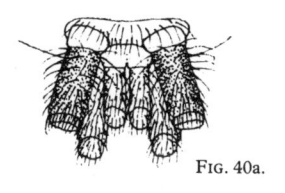

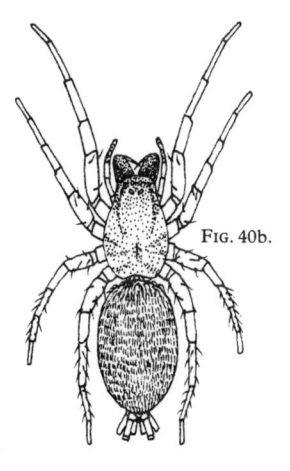

28a Labium much longer than broad (Fig. 41a), the front edge often markedly concave (Fig. 41b). Foot claws with 6–20 teeth and claw tufts (Fig. 41c) . Family CLUBIONIDAE

> **CLUBIONIDAE**
> The foliage spiders, with 23 British species in 2 genera, Clubiona and Cheiracanthium. They have elongate bodies, often without any defined pattern on the abdomen, although three species have an arrangement of chevrons (Fig. 41e). They are usually yellow-brown to dark brown, and the spinnerets are cone shaped (Fig. 41d). Nocturnal hunters, living in silken retreats during the day, they prey on a variety of other invertebrates. Mating takes place in the spring and egg-laying soon after, in June and July. The male grips the female with his chelicerae, but there is little courtship. The life cycle is probably completed in a year. Found in a variety of habitats, under stones and bark, in grass and other vegetation including the foliage of trees. Length: 3–10 mm.
> (LM&M Vol I p. 124, Vol III p. 12; Roberts Vol I p. 80, plts. 28–36)

28b Labium scarcely longer than broad, the front edge usually straight or slightly convex, rarely concave (Fig. 42). Foot claws with 0–5 teeth and without claw tufts (e.g. Figs. 43 & 44a) 29

29a All foot claws without teeth (Fig. 43). . . . .
Family GNAPHOSIDAE

> **GNAPHOSIDAE**
> The genus which keys out here, Phrurolithus with 2 species together with Micaria (which keys out under 27a) with 5, are in their own sub-family, the Micariinae, and may eventually be given full family status. The abdomen is dark with whitish markings and is iridescent in Micaria. When alive their movements, size and appearance are ant-like; although this is not always obvious in preserved specimens. Day-active species. Length: 2–5 mm.
> (LM&M Vol I p. 95, Vol II p. 406, Vol III p. 5; Roberts Vol I p. 64, plts. 20–27)

29b All foot claws with 3–5 teeth (Fig. 44a) . . . .
Family LIOCRANIDAE

> **LIOCRANIDAE**
> Closely allied to the clubionids, with 11 species in 4 genera, and called the running foliage spiders. They generally have markings on the carapace with a mottled, often indistinct abdominal pattern. Their bodies are elongate and they have relatively long legs (Fig. 44b). They are ground-running, mainly nocturnal species, capable of rapid progress, and feed on a variety of other invertebrates. They become adult in late summer and autumn, overwintering and appearing again in the spring when mating (which involves intense vibrating of the front pair of legs by the male) takes place. Eggs are laid in a remarkable, mud-daubed sac fixed high up on heather and grass stems during late May and June. The life cycle is probably completed in a year. They are found in wet and dry grasslands and heathlands. Length: 2.5–6 mm.
> (LM&M Vol I p. 146, as g. Agroeca, Agraecina, Scotina, Liocranum, Vol II p. 406, as f. Clubionidae, Vol III p. 15 as g. Agroeca, Agraecina, Scotina; Roberts Vol I p. 88, as sf. Liocraninae, plts. 32–36)

Keys to the Families of British Spiders

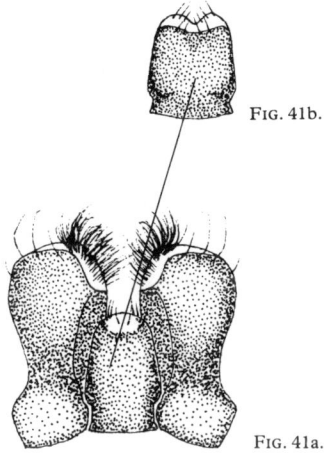

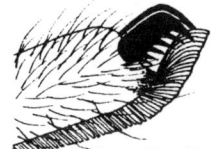

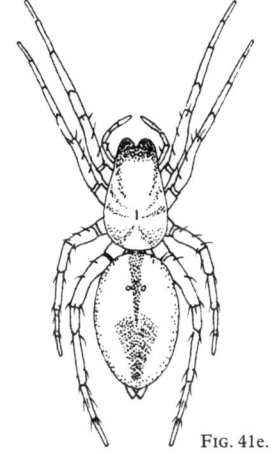

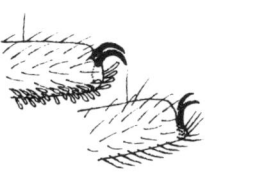

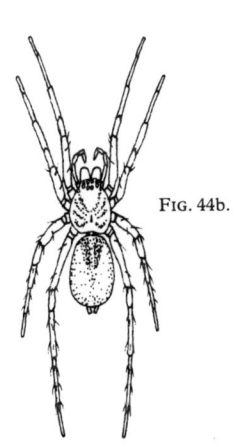

Fig. 41b.
Fig. 41c.
Fig. 41a.
Fig. 41d.
Fig. 41e.

Fig. 42.
Fig. 43.
Fig. 44a.
Fig. 44b.

TABLE 2

FAMILY	TARSAL COMB	MAXILLAE	LABIUM	TEETH ON FANG FURROWS	EYES	WEB
Nesticiidae	Present	Longer than broad	Rebordered	Present	Post. medians further apart than ant. medians	Irregular web
Theridiidae	Present (1)	Longer than broad	Flat	Absent (2)	Post. medians the same distance or further apart than ant. medians	Irregular web
Tetragnathidae	Absent	Longer than broad	Rebordered (3)	Present	Post. medians further apart than ant. medians	Orb web
Metidae	Absent	Longer than broad	Rebordered	Present	Post. medians the same distance or further apart than ant. medians	Orb web
Araneidae	Absent	As long or shorter than broad	Rebordered	Present	Post. medians closer together than ant. medians	Orb web
Theridiosomatidae	Absent	As long as broad	Rebordered	Present	Post. medians the same distance apart as ant. medians	Orb web (4)
Linyphiidae	Absent	Variable (5)	Rebordered	Present	Post. medians the same distance or further apart than ant. medians	Hammock/sheet web

Notes:
(1) The tarsal comb is often indistinct, especially in males of certain species, but when it is absent the combination of a flat labium and the absence of teeth on the fang furrows can be used as determining characteristics.
(2) When these are present, the tarsal comb is usually quite strong and easily visible. Teeth not obviously on the margins should be ignored.
(3) Not obviously rebordered in *Pachygnatha*, however, the distinctive appearance of the mouthparts is characteristic (refer to illustration in key).
(4) Characteristic orb web (refer to illustration in key).
(5) The maxillae in linyphiids are very variable in shape between species and often between sexes. Normally, they are shorter than broad, as long as broad, or only scarcely longer than broad.

Keys to the Families of British Spiders 411

The following group of families may give the beginner some problems owing to the similarity of certain families, and the difficulty in seeing some of the characters. Table 2 should provide some help if used in conjunction with the key and the glossary of terms.

30a   Tarsus IV provided with a ventral (lower surface) row of six to ten serrated bristles, forming a comb (e.g. Fig. 45b) . 31

30b   Tarsus IV without such a comb 32

31a   Labium with the edge thickened to form a definite border (rebordered) (Fig. 45a). Comb bristles are no longer than the bristles on the dorsal side of tarsus IV (Fig. 45b). . .
                                                 Family NESTICIDAE

NESTICIDAE

This family has a single British representative, Nesticus cellulanus, the comb-footed cellar spider. This species is very similar to a theridiid and even has a comb on the fourth tarsi. The cephalothorax is relatively small and there is a globular abdomen (Fig. 45c). Both are marked: the cephalothorax with a dark central stripe and dark margins, the abdomen with a variable, blotchy pattern. The legs are long and slender with few spines. The web is a loose platform with sticky threads attaching it to the surrounding stone or brick (Fig. 45d). These trap crawling insects which are then devoured. Adults are found during all seasons of the year suggesting that the life cycle may take longer than a year to complete. Mating is in early summer and egg sacs are produced between June and August. The male jerks the web several times and the female responds. The spider is found in caves, mines, cellars, drains and sewers, in deep shade among boulders in woodland and in grikes in limestone pavement. Length: 3–6 mm.
(LM&M Vol II p. 92, Vol III p. 60; Roberts Vol I p. 196, plt. 128)

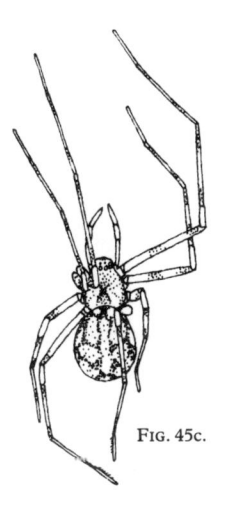

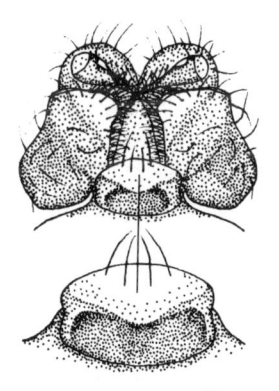

FIG. 45a.

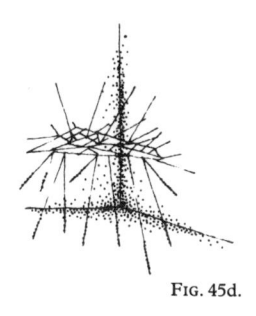

FIG. 45d.

FIG. 45b.

FIG. 45c.

31b Labium flat, with the edge not thickened to form a border (Fig. 46a). Comb bristles longer than the bristles on the dorsal side of tarsus IV (Fig. 46b). . . . .
Family THERIDIIDAE

THERIDIIDAE

The comb-footed spiders contain 52 British species grouped into 12 genera. This family contains the infamous black widow spider, although it is not a British species. Tarsus IV is characteristic, with a comb of six to ten serrated bristles, although this may be difficult to see in small species, and some adult males. The general appearance (but see note below) may vary between genera but a typical therid has a globular abdomen, a relatively small, often circular (when viewed from above) cephalothorax and slender legs with few spines (Fig. 46c).
NOTE: **Enoplognatha** *and* Robertus *with 11 species do not conform to this type; nor do* Steatoda *which have quite robust legs and* Episinius *which have a very angular abdomen. The abdomen often carries a distinctive pattern, although again there are exceptions, there are no teeth on the fang furrow and the labium is flat, without a swollen edge. One or two genera have species with strange and grotesque cephalothorax shapes. Webs vary from a complex, tangled mesh (Fig. 46d) to a few threads. A range of flying and crawling insects are taken, often many times larger than the spider itself. Some species specialise on eating ants, a group shunned by many spiders (but see* Zodariidae, *p. 393). Mating, which involves considerable vibrating and jerking of the female's web by the male prior to copulation, takes place in spring and summer and is followed soon after by the production of egg sacs. Some species may lay thousands of eggs, others like* Theridion sisyphium *restrict each brood to no more than thirty and feed the young, which cluster around the mouth, with regurgitated fluid. The full life cycle is usually completed in a year. They are found in a variety of habitats, occasionally running on the ground but usually in webs on vegetation, in wetlands, grasslands, heathlands, scrub and woodlands, in greenhouses and in buildings. Length: 1–10 mm.*
(LM&M Vol II p. 37, Vol III p. 47; Roberts Vol I p. 172, plts. 103–127)

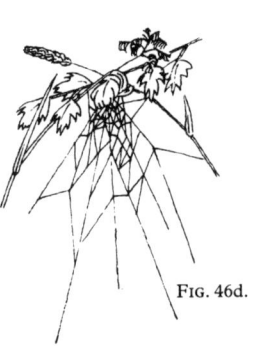

FIG. 46a.

FIG. 46b.

FIG. 46c.

FIG. 46d.

32a Maxillae much longer than broad (e.g. Figs. 47d &48c). 33
32b Maxillae shorter than, or only scarcely longer than broad (e.g. Fig. 54c) 34
Note: The shape is very variable in the Linyphiidae which key through couplet 34, but the mouthparts never resemble those illustrated in Figs. 47d, 47e & 48c.

NOTES
(key continues overleaf)

33a Epigastric furrow between lung slits is curved forwards (procurved) (Fig. 47a). . Family TETRAGNATHIDAE

TETRAGNATHIDAE
Called the long-jawed orb-weavers or, simply, long-jawed spiders this family contains 9 species in 2 genera, Tetragnatha *and* Pachygnatha. Tetragnatha *has a characteristically elongated abdomen with very long slender legs and large, powerful and divergent jaws in the adults (Fig. 47c). They weave orb-webs which are often inclined at an angle or even laid horizontally (Fig. 47f). Adults and late-instars of* Pachygnatha *are ground-running with relatively short legs and rounded abdomens (Fig. 47b). However, the very early instars have been shown to weave orb-webs. Several features are common to both* Pachygnatha *and* Tetragnatha*: the large, powerful jaws; distinctive trichobothria at the base of each femur; a procurved epigastric furrow; maxillae that are longer than broad (*Pachygnatha*), Fig. 47d; (*Tetragnatha*, Fig. 47e).* Tetragnatha *traps flying insects in its orb web.* Pachygnatha *actively hunts ground running prey. Mating occurs in early summer and eggs are laid soon afterwards. There is no courtship in either genus, the large jaws are simply locked together until copulation is complete. The full life cycle is probably completed in a year.* Tetragnatha *webs are usually found in tall vegetation, often in damp habitats, or the reeds around ponds.* Pachygnatha *is found running in grasslands, heathlands and on woodland floors. Length: 2.5–11 mm.*
(LM&M Vol II p. 94, Vol III p. 61; Roberts Vol I p. 198, plts. 129–135)

33b Epigastric furrow nearly straight (Fig. 48a) . . . . . Family METIDAE

METIDAE
This family of orb-weavers contains 8 species divided between 3 genera; these are Metellina, Meta *and* Zygiella *and they are closely related to the Araneidae. They all weave orb webs and have a rounded, patterned abdomen (e.g. Fig. 48b). The cephalothorax, which is normally pale, often has darker central markings. There are no trichobothria at the base of the femora. The maxillae are longer than broad (Fig. 48c). Flying, and occasional crawling, insects are trapped in the orb web (Fig. 48d), completely wrapped in silk and eaten. Mating, which involves a complicated series of vibrations and plucks of the female's web by the male (who may be in attendance for several weeks in* Metellina*), is in the summer and early autumn, after which eggs are laid. In some of the larger species the life cycle may take two or three years for completion. They are found in a variety of habitats:* Meta *in caves, mines, cellars, sewers, in deep shade in woodland and under bridges;* Metellina *on bushes and bramble;* Zygiella *outside buildings or on tree trunks. Length: 3.5–16 mm.*
(LM&M Vol II p. 111, as g. Meta, Zygiella, *Vol III p. 61, as g.* Meta*; Roberts Vol I p. 198, as g.* Meta, *plts. 133–135)*

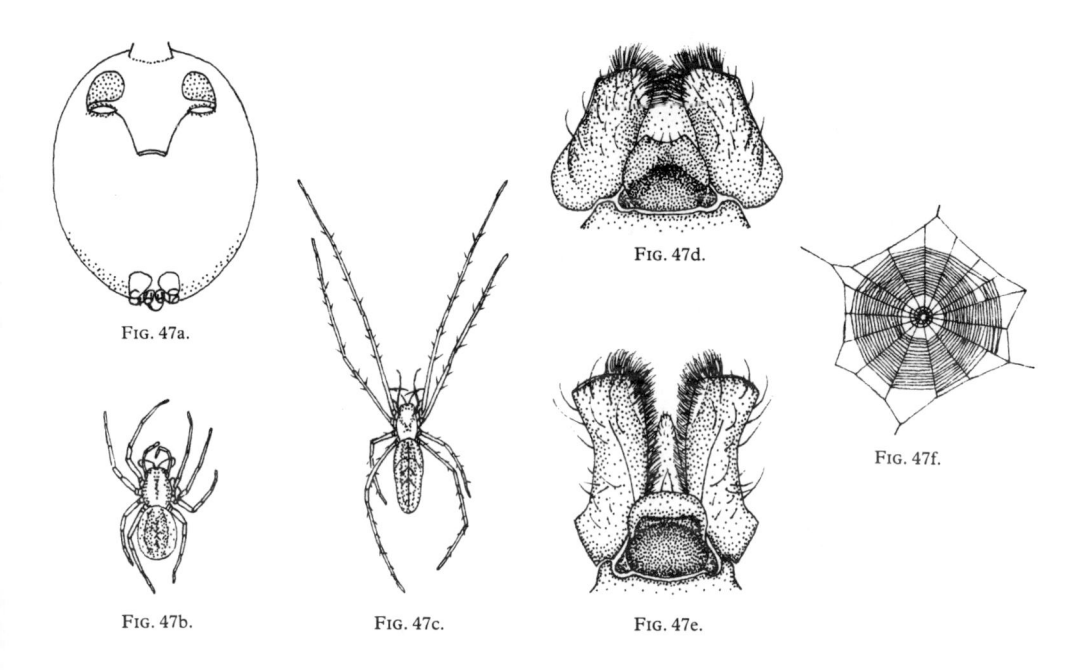

Keys to the Families of British Spiders

FIG. 47a.

FIG. 47b.

FIG. 47c.

FIG. 47d.

FIG. 47e.

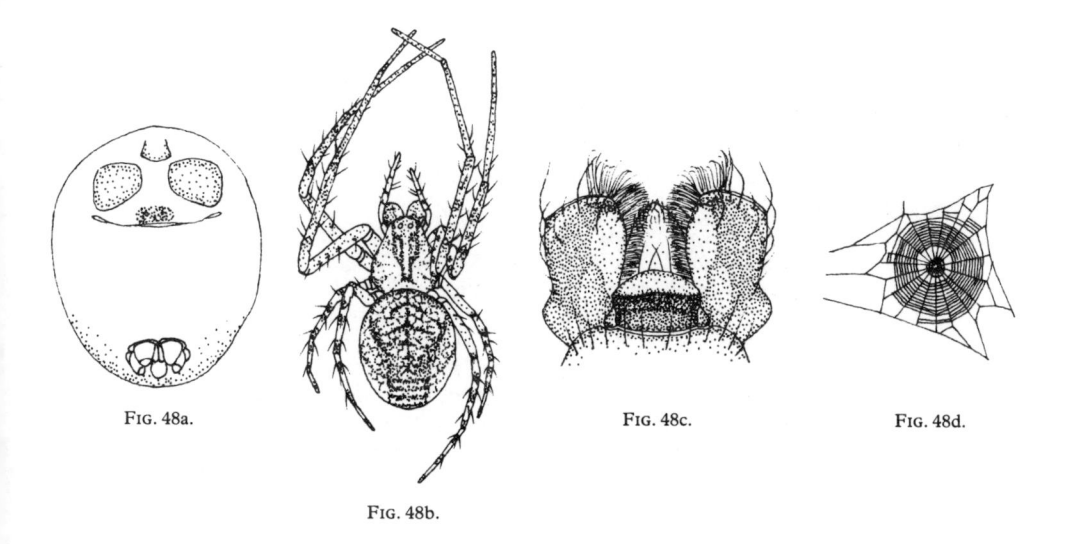

FIG. 47f.

FIG. 48a.

FIG. 48b.

FIG. 48c.

FIG. 48d.

416   L. M. JONES-WALTERS

34a   Abdomen globular and silvery (Fig. 49b). Tiny, adults never larger than 2 mm, femur I two to three times thicker than femur IV (best viewed from the side) (Fig. 49a) . . Family THERIDIOSOMATIDAE

THERIDIOSOMATIDAE
The only British species of this family is Theridiosoma gemmosum, *the Ray spider, named after the famous British naturalist John Ray. It is a tiny, globular spider with a distinctive silvery, patterned abdomen. The femora and tibia of the first leg are characteristically thickened and are between two and three times thicker than femur IV. The web is remarkable and sets the species apart from other orb weavers. The horizontal orb web is pulled into a taught conical shape by the spider who sits above the web at the end of a signal line attached to the centre (Fig. 49c). When a flying insect falls into the web it is released with a jerk, ensnaring the prey. In some ways this resembles the technique of the uloborid* Hyptiotes *(see p. 400). Mating is in the summer and the eggs are laid in a stalked sac hung in grassy vegetation. The life cycle is probably completed in a year and little is known of the mating ritual. The species is rare, being found low down in damp marshy vegetation. Length: 1.5–3 mm.*
(LM&M Vol II p. 168, as g. Theridiosoma; *Roberts Vol I p. 222, plt. 157)*

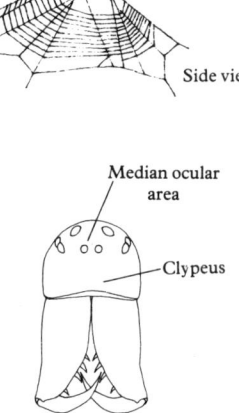

FIG. 49a.

FIG. 49b.

Top view

FIG. 49c.

Side view

34b   If less than 2 mm in length and abdomen globular, then not silvery. If femur I thicker than femur IV then spider not less than 2 mm in length 35

Note: The following group of families may give the beginner some problems. Table 2 should provide some help if used with the key and glossary of terms.

35a   When viewed from in front the clypeus is as high as or, more commonly, higher than the height of the median ocular area (Fig. 50) . . . . Family LINYPHIIDAE

LINYPHIIDAE
Known as the money spiders, this family contains 267 species in 105 genera—approximately half of the British spider fauna. They are generally small spiders and the strong teeth on the fang furrows, the rebordered labium, the lack of a cheliceral boss and the wide clypeus help to distinguish them. They are quite variable in form and it is useful to consider the two subfamilies, the Erigoninae and the Linyphiinae, separately. Foreign workers consider the Erigoninae to be a separate family, the MICRYPHANTIDAE, and the Linyphiinae to constitute the family LINYPHIIDAE. In fact, the relationships within the group are difficult to resolve and it may be some time before workers in Britain accept a separation of the group into two families. The two subfamilies are keyed out on p. 418.

Median ocular area

Clypeus

FIG. 50.

35b   When viewed from in front the height of the clypeus less than the height of the median ocular area (Fig. 51a,b) . 36

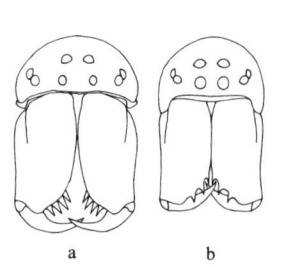

a   b

FIG. 51.

Keys to the Families of British Spiders

36a Anterior median eyes when viewed from in front much larger than the remainder, five to six large teeth on the front margin of the cheliceral fang furrow (Fig. 52) . . . . .
Family LINYPHIIDAE

A single species, *Tapinopa longidens*, keys out here. It is the only linyphiid, of over 250 species, to have a narrow clypeus. A full family description is given in couplet 35a, above.

FIG. 52.

36b Eyes all the same size and shape, teeth on the front margin of the cheliceral fang furrow 37

37a Head region contrastingly dark against pale thoracic region (Fig. 53a); orb web with a section missing (Fig. 53b)
Family METIDAE

Three species of the genus *Zygiella* key out here. They are very like the araneids in appearance but are considered to fall within the Metidae. A full description of the family is given in couplet 33b, above.

37b Head region not contrastingly dark against pale thoracic region; orb web complete (Fig. 54a) . . . .
Family ARANEIDAE

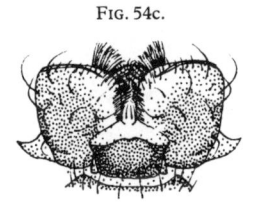

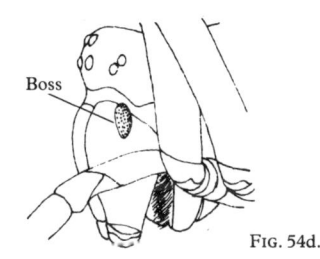

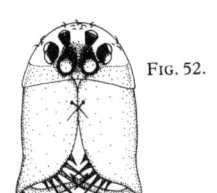

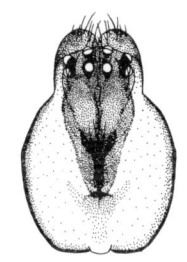

FIG. 53a.

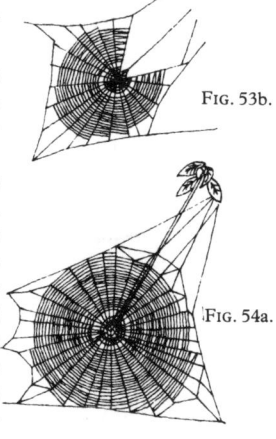

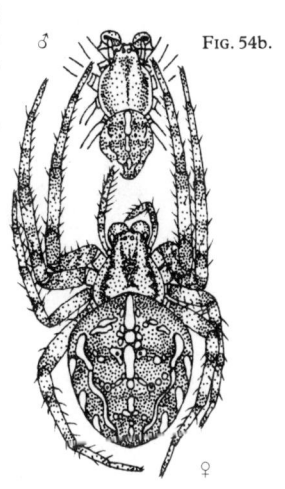

ARANEIDAE

Orb-weavers, containing 33 species in 15 genera. Adults have a large, patterned, globular abdomen that considerably overhangs the cephalothorax (Fig. 54b). The legs are often heavily spined, the maxillae are shorter than, or as long as, broad (Fig. 54c). There is a protrusion on the side of each chelicera called a boss (Fig. 54d). Flying insects are snared in their orb webs, wrapped with silk and eaten. Mating involves a complex series of vibration and jerking of the female's web by the male, who may mate with several females during the summer. Each time he risks loosing legs or palps until, finally, he cannot escape and at the last time of mating is devoured. Eggs are laid following mating and some of the larger species may take two or three years to reach maturity and complete their life cycle. The webs are found in a variety of habitats: spun horizontally in short grass in Hyposinga; *in long grass, heather, scrub and woodland in a variety of species; or in the eaves of buildings and across farm gates as in* Nuctenea umbratica. *This family includes the garden spider* Araneus diadematus. *Length: 3–15 mm.*
(LM&M Vol II p. 111, Vol III p. 64; Roberts Vol I p. 205, plts. 136–156)

FIG. 53b.

FIG. 54a.

FIG. 54b.

FIG. 54c.

Boss

FIG. 54d.

Key to sub-families of the family Linyphiidae

1a   Tibia IV with one dorsal spine and all metatarsi spineless . . .
Sub-family ERIGONINAE

FIG. 55a.

1b   Tibia IV with two dorsal spines, or if one spine is present, then there is a short spine on Metatarsi I and II . . . .
Sub-family LINYPHIINAE

***Erigoninae** (Micryphantidae)*
The dwarf, pigmy or midget spiders, this subfamily of the money spiders contains 148 species belonging to 58 genera. They are generally small, never larger than 3.5 mm, usually dark and often shiny spiders, without any abdominal patterning (Fig. 55a). The legs are usually short and are never heavily spined. The males of a small number of species have their heads modified into extraordinary, grotesque shapes (see Fig. 14), but the appearance of most species is otherwise unremarkable. The predatory behaviour is not well known and while a number of species in most genera weave small sheet webs these may be temporary structures made to create a favourable microclimate during periods of inactivity, not snares (Fig. 55b). Many erigonine species are probably ground running predators. Mating in some species involves the production of high frequency sound or vibration by the male, who moves a spur on the inside of his palps up and down stridulating ridges found on the sides of the chelicerae (Fig. 55c). During mating the female may grip the male by placing her fangs in the holes or pits sometimes found on the male's head, providing a partial explanation for the strange shape of the cephalothorax in some species. In many cases adults, particularly females, may be found all year. Male numbers are concentrated between the late summer and the spring. In the autumn, erigonines may be seen ballooning in huge numbers and they are active during the winter when snow is on the ground. Mating is probably between late summer and the spring and eggs are laid soon after; the full life cycle is completed in a year. Found in all habitats, in most abundance in low vegetation and leaf litter. Length: 1.2–3.5 mm.

FIG. 55b.

FIG. 55c.

FIG. 55d.

***Linyphiinae** (Linyphiidae)*
The line weaving spiders, this money spider subfamily contains 115 species divided into 46 genera. They often, though not always, have an abdominal pattern (e.g. Fig. 55d). The legs, which can be quite spiny, are long and slender relative to the body. The cephalothorax is usually unremarkable. Most species weave a horizontal sheet web which may be large and conspicuous in some genera, and this usually serves as a snare for flying or crawling insects which fall onto its upper surface (Fig. 55e). Mating takes place in late summer, with courtship in the web, and many males have stridulating ridges on the chelicerae. Eggs are layed soon afterwards and the life cycle takes a year to complete. The sub-family contains many ballooning species. Found in all habitats. Length: 1.2–6 mm.
(*LM&M* Vol II p. 172, Vol III p. 69; Roberts Vol II, plts. 158–237)

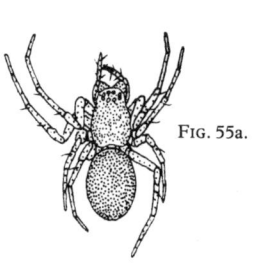

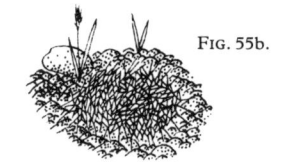

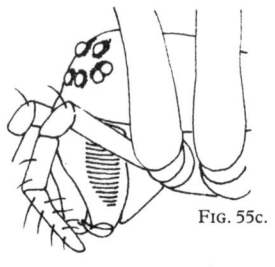

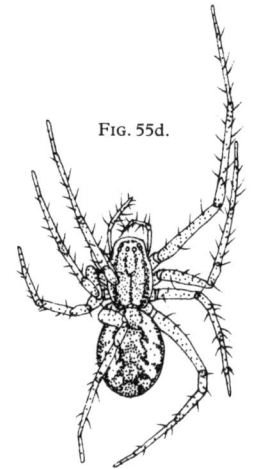

FIG. 55e.

NOTES
(Lateral key begins overleaf)

FAMILY (& details)	BODY FORM	CEPHALOTHORAX & EYES	LEGS/CLAWS
ATYPIDAE **Purse Web Spider** 1 species Couplet 1a 7–18 mm		Large, projecting chelicerae and eight eyes in distinctive pattern.	Three claws. Note distinctive rings on tarsus (r).
ERESIDAE **Ladybird spider** 1 species Couplet 2a Up to 15 mm	Female dark all over.	Eight eyes in distinctive pattern.	Legs very robust with calamistrum (c) on metatarsus IV. Three claws.
AMAUROBIIDAE **Lace-webbed spiders** 3 species Couplet 16a 4–15 mm		Eight eyes in two rows. Head area dark.	With calamistrum (c) on metatarsus IV of two rows in females. Three claws.

Keys to the Families of British Spiders

SPINNERETS	MOUTH PARTS	DISTINCTIVE FEATURES	WEB/ECOLOGY
One pair long and 3-segmented.	Chelicerae large and projecting.	Chelicerae. Two pairs of book lungs. Long, 3-segmented spinnerets. Purse web in field.	Closed tube half buried in soft soil. Males ground running Sept–Oct.
With cribellum (c).	Not distinctive.	Bodyform, eye pattern and cephalothorax, cribellum and calamistrum.	Buried tube and tangled sheet web in heather.
With cribellum (c).	Not distinctive.	Cribellum and calamistrum. Tarsi with dorsal row of trichobothria increasing in length towards tip (t). Web in field.	Tangled 'fuzzy' web with circular retreat in walls or tree bark. Occasionally found wandering in buildings.

FAMILY (& details)	BODY FORM	CEPHALOTHORAX & EYES	LEGS/CLAWS
DICTYNIDAE **Mesh-webbed spiders** 15 species Couplet 17b Up to 4 mm	Colour and pattern varies between species.	Eight eyes in two rows. Colour of cephalothorax varies between species.	With calamistrum (c) on metatarsus IV which may be weak and difficult to see in some males. Three claws.
ULOBORIDAE **Cribellate orb-weavers** 2 species Couplet 17a 3–6 mm Note: *Hyptiotes* above *Uloborus* below throughout.		Eight eyes in two rows.	Metatarsus IV compressed under line of calamistrum (c). Three claws.
OONOPIDAE **Six-eyed spiders** 2 species Couplet 10a Up to 2 mm		Six eyes in distinctive pattern.	Two claws on onychium (o). Note 'feathery' hairs.

SPINNERETS	MOUTH PARTS	DISTINCTIVE FEATURES	WEB/ECOLOGY
With cribellum (c).	Not distinctive.	Cribellum and calamistrum. Web in field.	Tangled 'fuzzy' web in heads of plants or across leaves.
With cribellum (c).	Not distinctive.	Cribellum and calamistrum. Cephalothorax form and eye pattern. Webs in field.	Triangular section of orb web. Full orb web strung horizontally with stabilimentum (s).
Not distinctive.	Not distinctive.	Eye pattern, onychium and pink colouration.	No web. Found in buildings, under bark, stones, in grass tussocks and occasionally in birds nests or the webs of other spiders. Nocturnal hunter and scavenger.

FAMILY (& details)	BODY FORM	CEPHALOTHORAX & EYES	LEGS/CLAWS
DYSDERIDAE **Six eyed spiders** 3 species Couplet 11a Up to 15 mm Note: When two illustrations *Harpactea* above *Dysdera* below.		Six eyes clustered in a circular pattern.	Three claws or two claws and claw tuft (c.t.). 3rd claw c.t.
SEGESTRIIDAE **Six-eyed spiders** 3 species Couplet 11b 6–22 mm		Six eyes in distinctive pattern.	Three claws. 3rd claw
SCYTODIDAE **Spitting spider** 1 species Couplet 9a 3–6 mm		Six eyes in distinctive pattern carapace patterned and domed in profile.	Legs long. Two claws with onychium (o). o

SPINNERETS	MOUTH PARTS	DISTINCTIVE FEATURES	WEB/ECOLOGY
Not distinctive.	Not distinctive.	Eye pattern. Jaws and deep red cephalothorax and pink abdomen in Dysdera.	No web but build a silken retreat under stones. Nocturnal hunters.
Not distinctive.	Not distinctive.	Eye pattern, abdominal markings, elongate body form with first three pairs of legs thrust forwards.	A silk tube in walls, under bark or stones with up to 12 distinctive threads radiating from the lip.
Not distinctive.	Modified, but not obviously distinctive. Fangs pincer-like.	Eye pattern, body form and pattern.	No web. Wander over walls and ceilings in buildings. Catch prey by spitting gummy threads from chelicerae.

FAMILY (& details)	BODY FORM	CEPHALOTHORAX & EYES	LEGS/CLAWS
PHOLCIDAE **Cellar or Daddy-long-legs spiders** 2 species Couplet 7a 2–10 mm		Eight eyes in two rows with distinctive arrangement. Cephalothorax almost circular.	Legs extremely long and slender. Tarsi flexible, apparently segmented (s) under certain light. Three claws. s 3rd claw
ZODARIIDAE **Ant-eating spiders** 1 species Couplet 6a 2–3 mm		Eight eyes in distinctive pattern, the anterior medians large.	Three claws. 3rd claw
GNAPHOSIDAE **Ground spiders** 33 species Couplets 27a & 29a 2–18 mm		Eight in two rows. Posterior medians oblique, oval, triangular.	Two claws with or without claw tufts (c.t.). c.t

SPINNERETS	MOUTH PARTS	DISTINCTIVE FEATURES	WEB/ECOLOGY
Not distinctive.	Pincer like fangs. Otherwise not distinctive.	Legs and body form. Eye pattern and almost circular cephalothorax.	Tangled, open web in buildings. Occasionally leave webs to prey on other spider species.
Apparently with only a single pair of tube-like spinnerets.	Not distinctive.	Spinnerets and eye pattern. Abdomen dark brown above and contrastingly pale yellow below.	No web. Found running with black ants (*Lasius niger*) on which they exclusively feed.
Upper pair cylindrical and separated at their bases by a distance equal to their diameter. More heavily sclerotised than lower ones.	Not distinctive.	Eyes, spinnerets and claws. 'Sleek' appearance of abdomen due to short dense hairs.	No web. Ground running, nocturnal hunters which feed mainly on other spiders.

FAMILY (& details)	BODY FORM	CEPHALOTHORAX & EYES	LEGS/CLAWS
CLUBIONIDAE **Foliage spiders** 23 species Couplet 28a 3–10 mm		Eight eyes in two rows.	Two claws with claw tufts (c.t.). Foot claws with 6–20 teeth.
LIOCRANIDAE **Running foliage spiders** 11 species Couplet 29b 2.5–6 mm		Eight eyes in two rows.	Two claws without claw tufts. Foot claws with 3–5 teeth.
ZORIDAE **Ghost Spiders** 4 species Couplet 13a 2.5–6.5 mm		Eight eyes in three rows.	Two claws with claw tufts (c.t.)

SPINNERETS	MOUTH PARTS	DISTINCTIVE FEATURES	WEB/ECOLOGY
Cone shaped, hardly separated at their bases.	Labium longer than broad, front edge often markedly concave.	Spinnerets and labium, claws.	No web. Nocturnal hunters, living in silken retreats during the day, found on foliage and grass.
Cone shaped, hardly separated at their bases	Labium scarcely longer than broad the front edge usually straight or slightly convex	Spinnerets and labium. Claws. *Note:* the gnaphosid *Phrurolithus* will 'fit' here. The claws, however, lack any teeth and the spider has a black abdomen with white markings.	Ground running, mainly nocturnal species capable of rapid progress. Hunt other invertebrates.
Not distinctive.	Not distinctive.	Eye pattern and narrow head area. Pale background colour with darker markings.	No web. Hunting species, often found during the day on ground vegetation.

FAMILY (& details)	BODY FORM	CEPHALOTHORAX & EYES	LEGS/CLAWS
ANYPHAENIDAE **Buzzing spider** 1 species Couplet 24a 4–7.5 mm		Eight eyes in two rows.	Two claws with claw tufts (c.t.).
EUSPARASSIDAE **Green spider** 1 species Couplet 20a 7–13 mm	Legs I & II turned forwards.	Eight eyes in two rows ringed with white hairs (which fade in preserved specimens), black and beady.	Two claws with claw tufts. Tarsi and metatarsi I and II with dense scopuli (s).
THOMISIDAE **Crab spiders** 25 species Couplet 19a 2–10 mm	Crab-like. Legs I & II turned forwards.	Eight eyes, in two rows, black and beady. Laterals often on tubercles. Cephalothorax almost circular.	Two claws without claw tufts. Tarsi and metatarsi without scopuli.

SPINNERETS	MOUTH PARTS	DISTINCTIVE FEATURES	WEB/ECOLOGY
Not distinctive.	Not distinctive.	Abdominal pattern. Tracheal slit (t.s.) midway between spinnerets and epigastric fold.	No web. Hunts over foliage of trees and bushes. Rarely found on the ground.
Not distinctive.	Rear margin of fang furrow armed with teeth. 'Fan' of stiff hairs on face of chelicerae (f).	Colouration: bright green in adult females; green with red and yellow abdomen in males. White rings around eyes, eye colour, tarsal and metatarsal scopuli and cheliceral features. Note: Colour fades to uniform yellow in preserved specimens.	No web. Hunt in grassy vegetation, more often in the south.
Not distinctive.	Not distinctive.	'Crab-like' appearance with legs I and II turned forwards, an almost circular cephalothorax and a dumpy, squat abdomen.	No web. Lie in wait for prey on the ground, on vegetation and in flower cups and are often cryptically coloured.

FAMILY (& details)	BODY FORM	CEPHALOTHORAX & EYES	LEGS/CLAWS
PHILODROMIDAE **Running crab spiders** 15 species Couplet 20b 4–7 mm	Legs I and II turned forwards	Eight eyes in two rows, black and beady. Cephalothorax almost circular.	Two claws with claw tufts. Tarsi and metatarsi I and II with scopuli (s).
SALTICIDAE **Jumping spiders** 34 species Couplet 3a 2–10 mm	Squat and compact.	Eight eyes in three rows. Front row big, facing forwards.	Two claws with claw tufts (c.t.). Legs short.
OXYOPIDAE **Lynx spider** 1 species Couplet 5a 5–8 mm	Legs I–III characteristically drawn back over cephalothorax.	Eight eyes in hexagonal pattern.	Three claws. Legs very spiny.

SPINNERETS	MOUTH PARTS	DISTINCTIVE FEATURES	WEB/ECOLOGY
Not distinctive.	Rear margin of fang furrow without teeth.	Body form, with legs I and II turned forwards. Black, beady eyes, almost circular cephalothorax. Tarsal and metatarsal scopuli.	No web. Active hunters in foliage and ground vegetation, occasionally in buildings. Capable of rapid movement.
Not distinctive.	Not usually distinctive although chelicerae of some males may project forwards.	Eyes and compact appearance.	No web. Hunt on foliage and ground vegetation, often on sunny walls. Jump onto prey.
Not distinctive.	Not distinctive.	Hexagonal eye pattern, spiny legs, markings on cephalothorax and abdomen. Stance, in life, with femora I–III drawn back over cephalothorax.	No web. Hunt in heather.

FAMILY (& details)	BODY FORM	CEPHALOTHORAX & EYES	LEGS/CLAWS
LYCOSIDAE **Wolf spiders** 36 species Couplet 14a 3.5–18 mm		Eight eyes in three rows.	Three claws. 3rd claw
PISAURIDAE **Nursery web and raft spiders** 3 species Couplet 14b 9–22 mm Note: *Dolomedes* above, *Pisaura* below in first box.		Eight eyes in three rows.	Three claws. 3rd claw
ARGYRONETIDAE **Water spider** 1 species Couplet 25a 8–15 mm		Eight eyes in two rows	Three claws. Legs III and IV covered with dense hairs. 3rd claw

SPINNERETS	MOUTH PARTS	DISTINCTIVE FEATURES	WEB/ECOLOGY
Not distinctive.	Not distinctive.	Eye pattern. Egg sac attached to spinnerets in field.	No web, although some species live in a silken tube. Usually ground running hunters. Females carry egg sacs attached to their spinnerets.
Not distinctive.	Not distinctive.	Eye pattern. Large size. Body markings. Egg sac carried in jaws in field. Nursery webs.	Ground running hunters, *Dolomedes* also able to take aquatic prey by moving across still water. Both genera construct 'nursery' webs for their young. Females carry egg sacs in jaws.
Not distinctive.	Not distinctive.	Tracheal slit (t.s.) behind epigastric fold. Legs. Habitat.	Bell web constructed under water. Completely aquatic, living and hunting in slow flowing or still water.

FAMILY (& details)	BODY FORM	CEPHALOTHORAX & EYES	LEGS/CLAWS
AGELENIDAE **Funnel-web or cobweb spiders** 17 species Couplet 23b 2–17 mm		Eight in two rows.	Three claws. Tarsi with dorsal row of trichobothria increasing in length towards tip (t). 3rd claw
HAHNIIDAE **Lesser cobweb spiders** 7 species Couplet 23a 1.3–3 mm		Eight eyes in two rows	Three claws 3rd claw
MIMETIDAE **Pirate spiders** 4 species Couplet 21a 2.4–4 mm		Eight eyes in two rows.	Legs I and II with distinctive spine pattern. 3 claws.

SPINNERETS	MOUTH PARTS	DISTINCTIVE FEATURES	WEB/ECOLOGY
One pair (s) two segmented.	Not distinctive.	Spinnerets and tarsal trichobothria. Web in field.	Cobweb. In buildings and ground vegetation.
Arranged in a row, the outer pair two-segmented (s).	Not distinctive.	Spinnerets.	Small sheet web in soil, moss, leaf litter.
Not distinctive.	Not distinctive.	Spination of legs I and II. Body form.	Only spin a temporary web. Prey on other spiders by entering their webs and mimicking prey struggles.

FAMILY (& details)	BODY FORM	CEPHALOTHORAX & EYES	LEGS/CLAWS
THERIDIIDAE **Comb-footed spiders** 52 species Couplet 31b 1–10 mm		Eight eyes in two rows.	Tarsus IV with a comb (c). Three claws.
NESTICIDAE **Comb-footed cellar spider** 1 species Couplet 31a 3–6 mm		Eight eyes in two rows.	Tarsus IV with a comb (c). Three claws.
TETRA- GNATHIDAE **Long-jawed orb weavers, long-jawed spiders** 9 species Couplet 33a 2.5–11 mm Note: Where two illustrations *Pachygnatha* above *Tetragnatha* below.		Eight eyes in two rows.	Three claws. 3rd claw 3rd claw

SPINNERETS	MOUTH PARTS	DISTINCTIVE FEATURES	WEB/ECOLOGY
Not distinctive.	Labium flat. Fang furrows without teeth.	Tarsus IV. Labium and fang furrow. Globular body form.	Irregular tangle of criss-cross threads in a range of situations.
Not distinctive.	Labium rebordered. Fang furrows with teeth.	Tarsus IV. Labium. Long legs.	A loose platform with sticky threads attaching it to stone or brick. In cellars, caves, deep shade.
Not distinctive.	Maxillae longer than broad.	Epigastric furrow (e.f.) curved forwards. Chelicerae. Mouthparts with maxillae longer than broad. Web in field.	Orb web with open hub, often in damp, marshy vegetation. Usually strung at an angle or horizontally. *Pachygnatha* ground running, without web.

FAMILY (& details)	BODY FORM	CEPHALOTHORAX & EYES	LEGS/CLAWS
METIDAE **Orb weavers** 8 species Couplet 33b 3.5–16 mm		Eight eyes in two rows. Posterior medians (p.m.) same distance or further apart than anterior medians.	Three claws. 3rd claw
ARANEIDAE **Orb weavers** 33 species Couplet 37b 3–15mm		Eight eyes in two rows. Posterior medians (p.m.) closer together than anterior medians.	Three claws. Legs heavily spined. 3rd claw
THERIDIOSO-MATIDAE **Ray spider** 1 species Couplet 34a 1.5–3 mm	Abdomen silver.	Eight eyes in two rows. Posterior medians some distance apart as anterior medians.	Three claws. Femur I 2–3 × thicker than femur IV. 3rd claw

Keys to the Families of British Spiders

SPINNERETS	MOUTH PARTS	DISTINCTIVE FEATURES	WEB/ECOLOGY
Not distinctive.	Maxillae longer than broad. (Except in *Zygiella*—for which note eyes, web and dark head area.)	Epigastric furrow (e.f.) straight mouthparts with maxillae longer than broad. Eye pattern. Web in field.	Orb web with open hub or, in *Zygiella*, with section absent. Usually strung vertically.
Not distinctive.	Chelicerae with boss (b). Maxillae not longer than broad.	Maxillae not longer than broad. Chelicerae boss. Eye pattern. Abdomen overhangs cephalothorax. Web in field.	Orb web with closed hub. Found in a variety of habitats.
Not distinctive.	Maxillae not longer than broad.	Femoral width, I to IV. Silvery, globular abdomen. Size. Web in field.	Horizontal 'umbrella' shaped orb web, in damp, marshy vegetation.

FAMILY (& details)	BODY FORM	CEPHALOTHORAX & EYES	LEGS/CLAWS
LINYPHIIDAE **Money spiders** 267 species Couplets 4a, 35a & 36a 1.2–6 mm Note: subfamily *Erigoninae* above, *Linyphiidae* below, in first box.		Eight in two rows. Clypeus wide. Except in some males which have lobes or globular extensions to head area (see Couplet 4 dicot. key).	Three claws, variable pattern.

SPINNERETS	MOUTH PARTS	DISTINCTIVE FEATURES	WEB/ECOLOGY
Not distinctive.	Maxillae variable. Not longer than broad (include first basal width). Lack cheliceral boss.	Size. Head shape in certain males (dicot. key Couplet 4). Chelicerae in some species with stridulating ridges (s.r.). Web in field.	Sheet web on ground or in vegetation. Some species highly dispersive, rarely if ever in webs, found wandering or ballooning.

AIDGAP PUBLICATIONS

The Field Studies Council (FSC) will have published fifteen AIDGAP keys by the end of 1989:—

Hiscock, Sue (1979). *A field key to the British brown seaweeds* (FSC Publication 125).

Sykes, J. B. (1981). *An illustrated guide to the diatoms of British coastal plankton* (FSC Publication 140).

Unwin, D. M. (1981). *A key to families of British Diptera* (FSC Publication 143).

Crothers, John & Marilyn (1983). *A key to the crabs and crab-like animals of British inshore waters* (FSC Publication 155).

Cameron, R. A. D., Eversham, B. & Jackson, N. (1983). *A field guide to the slugs of the British Isles* (FSC Publication 156).

Unwin, D. M. (1984). *A key to the families of British Coleoptera and Strepsiptera* (FSC Publication 166).

Willmer, Pat (1985). *Bees, ants and wasps—the British Aculeates* (FSC Occasional Publication 7).

Pankhurst, R. J. and Allinson, J. (1985). *British Grasses: a punched-card key to grasses in the vegetative state* (FSC Occasional Publication 10).

King, P. (1986). *Sea Spiders: a revised key to the adults of littoral pycnogonida of the British Isles* (FSC Publication 180).

Croft, P. (1986). *A key to the major groups of British Freshwater Invertebrates* (FSC Publication 181).

Hiscock, Sue. (1986). *A field guide to the British Red Seaweeds* (FSC Occasional Publication 13).

Tilling, S. M. (1987). *A key to the major groups of British Terrestrial Invertebrates* (FSC Publication 187).

Friday, L. E. (1988). *A key to adult British water beetles* (FSC Publication 189).

Trudgill, Stephen (1989). *Soil types: a field identification guide* (FSC Publication 196).

Jones-Walters, L. M. (1989). *Keys to the families of British Spiders* (FSC Publication 197).

Another key was published before the AIDGAP project was initiated, but it has been fully tested and revised:—

Haslam, S. M., Sinker, C. A. & Wolseley, P. A. (1975). *British Water Plants* (FSC Publication 107).

These, and many other titles, may be purchased when visiting Field Studies Council Centres or may be ordered through the post from:—

FSC Publications, Field Studies Council, Central Services, Preston Montford, Montford Bridge, Shrewsbury SY4 1HW,
or from
The Richmond Publishing Company Ltd., P.O. Box 963, Slough SL2 3RS.

A complete list of titles and prices is available from either of these addresses.